D0992055

Probability Distributions on Linear Spaces

NORTH HOLLAND SERIES IN

Probability and Applied Mathematics

A. T. Bharucha-Reid, *Editor*

AN INTRODUCTION TO STOCHASTIC PROCESSES
D. Kannan

STOCHASTIC METHODS IN QUANTUM MECHANICS
S. P. Gudder

APPROXIMATE SOLUTION OF RANDOM EQUATIONS
A. T. Bharucha-Reid (ed.)

PROBABILITY DISTRIBUTIONS ON LINEAR SPACES
N. N. Vakhania

Probability Distributions on Linear Spaces

N. N. VAKHANIA

Computing Center
Academy of Sciences of the Georgian SSR, Tbilisi
and
Department of the Theory of Stochastic Processes
Tbilisi State University
Tbilisi, U.S.S.R.

NORTH HOLLAND
New York • Oxford

Elsevier North Holland, Inc.
52 Vanderbilt Avenue, New York, New York 10017

Sole distributors outside the USA and Canada:

Elsevier Science Publishers B.V.
P. O. Box 211, 1000 AE Amsterdam, The Netherlands

Library of Congress Cataloging in Publication Data

Vakhania, N. N. (Nikolaĭ Nikolaevich)
Probability distributions on linear spaces.
(North Holland series in probability and applied mathematics)
Translation of: Veroi͡atnostnye raspredelenii͡a v lineĭnykh prostranstvakh.
Bibliography: p.
Includes indexes.
1. Distribution (Probability theory) 2. Linear topological spaces.
I. Title. II. Series.
QA273.43.V3413 519.2 81-5543
ISBN 0-444-00577-3 AACR2

Desk Editor Philip Schafer
Design Edmée Froment
Design Editor Glen Burris
Production Manager Joanne Jay
Compositor Computype, Inc.
Printer Haddon Craftsmen

Manufactured in the United States of America

Contents

Editor's Preface to the English Edition

During the past twenty-five to thirty years the theory of probability in Banach and other linear topological spaces has developed very rapidly, and in a systematic fashion. One of the most important areas of research is concerned with the study of probability distributions (or measures) on Banach spaces. This English edition of Professor Vakhania's Russian text is the first systematic and clearly written introduction to the theory of probability distributions on linear spaces. This book can be used as a class or seminar text, for self-study, and as a reference work. It will be of interest to a broad audience of probabilists, functional analysts, measure theorists, statisticians, and a wide range of applied mathematicians who need results on probability distribution on linear spaces in connection with research on random equations and related topics.

Results presented in this book are extended in the following two papers by Professor Vakhania and his students: (1) S. A. Chobanyan and V. I. Tarieladez, "Gaussian characterization of certain Banach spaces." *J. Multivar. Anal.* 7: 183–203, 1977, and (2) N. N. Vakhania and V. I. Tarieladez, "Covariance operators of probability measures in locally convex spaces," *Theory Probab. Appl.* 23: 1–21, 1978. The reader is encouraged to study the above papers and to investigate the recent literature on the connections between probability theory in Banach spaces and their geometric properties.

Professor Vakhania has made some minor changes and corrections to this translation of the Russian edition by Professor I. I. Kotlarski. The series editor has made some changes in style.

A. T. Bharucha-Reid
Series Editor

Introduction

Probability distributions on linear spaces arise in probability theory in the following way. In the classical one-dimensional case, instead of investigating random variables, one can investigate probability distributions on the real line. According to this, in the infinite-dimensional case, for the process $\xi_t (t \in T)$, one should consider distributions on the family of all sample functions, that is, on the space R^T of all real functions on T. But it is only in the finite-dimensional case (when T is a finite set), that the analogy is sufficiently good. In the infinite-dimensional case problems arise that make little sense or are trivial in the finite-dimensional case. In reality, something known in advance about the random vector in R^n, does not necessarily help to simplify the sample space. For example, if it is known that the distribution is concentrated on some subspace, then we have again R^k, where $k < n$, but this is no simplification. In the case of a function space R^T, a priori information that the sample functions belong to some linear subspace can be used because different linear subspaces can have different natural topologies (which would not make sense for the whole R^T). This allows different additional possibilities of describing the process under consideration. Sometimes the same process can be embedded into different subspaces, thus varying the methods of investigation and the terms in which the process is described.

In this way, as a natural abstraction, there arises the notion of a probability distribution on an abstract linear space—in other words, the notion of a random element with values in a linear space.

There exists a different approach to these notions, which is disassociated from probability theory. This approach, which comes from the theory of functions and functional analysis, is motivated by the inner logic of developments in pure analysis. It is quite natural to extend the theory of integration to include functions defined on infinite-dimensional linear spaces, as well as functions of one and several variables. This endeavor immediately brings us to the notion of measures on linear spaces. Finite measures, normed by one, are called *probability measures*, or *probability distributions*.

In general, the problem of investigation of normed measures on linear spaces is very broad and has many different directions with or without emphasizing the possibility of probabilistic interpretation. Let us note the basic results of some of the directions, on which this book is based. Although all these investigations will not be covered, we shall go through a short description of the results that are given and therefore start with some background information.

In 1935 A. N. Kolmogorov [1] introduced the notion of a *characteristic functional* (i.e., of the Fourier transform) of the measure μ on the Banach space X (as an integral with respect to μ of $\exp[if(x)]$, $f \in X^*$). He gave the basic properties of the characteristic functional and pointed out the importance and possibility of further investigations in this area.

In 1951 M. Fréchet [2] examined Gaussian random elements (Gaussian distributions) in a Banach space. The random element $x \in X$ was called *Gaussian* if all real-valued random variables $f(x)$, $f \in X^*$, were Gaussian.

A systematic investigation of random elements with values in Banach spaces was begun in the works of E. Mourier and R. Fortet. In her basic work E. Mourier [3], in 1953, defined the expectation, using the weak integral (*Pettis integral*) of a random element. Further, for the case where X is a separable Hilbert space H, she defined an analogue of the variance, namely, the *covariance operator*, presupposing the existence of the expectation of the square of the norm of the random element. For Gaussian random elements in H, Mourier proved the existence of the expectation itself and of the expectation of the square of the norm, thus giving one of her basic results—the general form of the characteristic functionals of all Gaussian distributions on H.

The next steps in the investigation of distributions on Hilbert spaces are found in the works of Yu. V. Prohorov and V. V. Sazonov. In 1956 Yu. V. Prohorov [4] gave the analogue of Lévy's distance between distribution functions for the case of measures in metric spaces, and investigated the conditions for compactness and convergence of families of measures relative to this metric. Applying these results to distributions on Hilbert spaces, Prohorov obtained theorems expressing conditions for relative compactness of a family of distributions in terms of the corresponding family of characteristic functionals. These results served as the basis for a further result given by V. V. Sazonov [5] in 1958. He found that with respect to the topology on H, the analogue of Bochner's theorem is true: *A positive-definite normed complex functional on H is a characteristic functional of some distribution on H if and only if it is continuous with respect to this topology.*

These problems are closely related to the problem of extension of a weak distribution to a measure. While solving this problem, R. A. Minlos [6] independently obtained a theorem about the extension of generalized random processes, from which the analogue of Bochner's theorem follows for the distributions on spaces which are conjugate to countably normed Hilbertian nuclear spaces. The connection between the results of R. A. Minlos and V. V. Sazonov was perceived and analyzed by A. N. Kolmogorov [7].

Further references to other works connected to these problems will be given later, generally in the introductions to the chapters.

We will now briefly describe the book.

It contains four chapters. In the first chapter the general problems of probability distributions on the linear topological space R^N of all real numerical sequences with the usual linear operations and Tikhonov's topology are studied. After a preliminary description of the space R^N and its conjugate, we prove the analogues of the theorems of Bochner and Lévy for characteristic functionals.

The space R^N is first of all the natural sample space for random sequences. It is also interesting because a large class of linear spaces X may be embedded linearly and isomorphically into R^N, which allows reduction of problems from X to R^N. This process is particularly effective if X is a Hilbert space, because in this case its image is l_2, and the isomorphism is isometric (*Riesz–Fischer theorem*). Therefore all results proved for l_2 are also true for a general (separable) Hilbert space.

Further, we consider probability distributions on the Banach spaces $l_p(1 \leqslant p < \infty)$, c_0, and l_∞. We obtain some results analogous to Bochner's theorem under some conditions on the moments. Moreover, conditions of relative compactness of families of distributions on these spaces are investigated, along with the relations of these conditions to the properties of the corresponding families of characteristic functionals. Thus a result analogous to Lévy's theorem is obtained, which is the best possible in this situation. The basic idea in getting these results is the following: Each of the spaces l_p, c_0, l_∞, is considered as a linear subspace of R^N, and the theorem proved for R^N is applied. In this way the problem is reduced to obtaining conditions under which the distribution on R^N will be concentrated on a given subspace of R^N. The idea of enlarging the space is not, of course, new. However, we did not limit ourselves to considerations in the whole space. This space is a helping tool only and is not involved in the final formulations.

The method of embedding in R^N is also used in the second chapter, in which we investigate Gaussian distributions on R^N and its subspaces. First, the case of the whole space R^N is considered, and then we go to l_p, $1 \leqslant p < \infty$. After proving some auxiliary theorems we come to the basic theorem of this chapter: *In order that a Gaussian distribution R^N with* *expectation* $\{a_k\}$ *and covariance matrix* $\|s_{ij}\|$ *be concentrated on* l_p *(and hence to be Gaussian in* l_p*), it is necessary and sufficient that* $\{a_k\} \in l_p$ *and* $\{s_{kk}\} \in l_{p/2}$.

Then, we give two other formulations of this result. One is a theorem on the general form of characteristic functionals of all Gaussian distributions on l_p, which sheds new light on and is a generalization of the aforementioned result of E. Mourier. The other formulation gives an analogue of Bochner's theorem for Gaussian distributions on l_p by constructing a proper topology in the conjugate space. This topology is a natural generalization of Sazonov's topology.

Further, we show some other facts about Gaussian distributions on l_p, which are connected with different aspects of the theory of stochastic processes and their applications. By doing this we can consider these results as corollaries from the general theorems already proved, or as examples that illustrate their applications. We mention the proof of the exponential integrability of the square of the norm with respect to the Gaussian distribution, the central limit theorem that contains Mourier's theorem ($p = 2$) and Varadarajan's theorem ($p = 1$), the investigation of a stochastic differential equation of heat conduction with white noise on the right side (the Fourier transform of the solution of this equation determines the Gaussian distribution on R^N), and the characterization of nondegenerate Gaussian distributions (the nondegeneracy of μ is equivalent to the following property: The μ-measure of an arbitrary ball with a nonzero radius is positive).

Finally, in this chapter we investigate Gaussian distributions on the spaces c_0 and l_∞. We shall mention here only one result: For each $R > 0$, there exists on l_∞ a nondegenerate Gaussian distribution μ, such that the μ-measure of an arbitrary ball with radius R is zero.

The third chapter considers a separable Hilbert space H. Based on previous comments, the space H can be treated as l_2, hence we can use at once the results of previous theorems for this case. In this way we again obtain the known, as well as some new results.

Next, we come to some special problems, which are quite usual and traditional for the theory of probability. This part of the chapter is also connected with the methods and results of the second chapter. First, we consider the question of evaluating the rate of convergence in the central limit theorem. We consider the finite-dimensional case separately; and, in particular, when the coordinates of the vectors are independent, we obtain a uniform estimate in the class of all balls with the classical order for n (namely, $n^{-1/2}$), using the sum of Lyapunov's ratios as a parameter, which determines the dependence on the distribution. Then, we obtain in the case of a general Hilbert space, a uniform estimation of order $(\ln n)^{-1}$ on the class of ellipsoids with centers from a given ball, by operating on the distributions and applying the characteristic functionals.

Then, we consider the problem of evaluating large deviations for normed sums of independent random elements in H with bounded norms. An exponential type inequality is obtained for the evaluation of the measures of the complements of the ellipsoids that are defined by the nuclear operators.

Next in the third chapter we consider the distribution of the inner product of two Gaussian random elements in H. This problem is connected with the eigenvalue problem for a system of two operator equations. Particular cases are shown in which the solution is found in one form or another.

Finally, we define and give some simple properties of an integral with respect to a random measure with values in H, of a function whose values are linear operators $H \to H$. This integral is a random element in H; and we find, in particular, its covariance operator, which is in the form of a (nonrandom) integral in the strong sense (*Bochner's integral*).

The fourth chapter is devoted to some general problems of probability distributions on abstract Banach spaces. We discuss the notion of a characteristic functional and note its basic properties. We prove a theorem on conditions for the existence of the Pettis integral, which includes a theorem on the existence of the expectation of an arbitrary Gaussian distribution on a separable space X. Then, the *covariance operator* R is defined as a bounded linear transformation $X \to X^{**}$. The conditions for the existence of R are reduced to the natural necessary conditions; in particular, each Gaussian distribution has a covariance. The operators that transform a space into its conjugate have special properties, which distinguish them from the general transformations of one Banach space into another. For instance, the notions of symmetry and nonnegativeness can be applied to such operators, and all covariance operators have these properties. Under some broad assumptions the converse theorem is proved: *Every symmetric, nonnegative, bounded linear operator* $X^* \to X^{**}$ *is the covariance operator of some distribution on* X. The proof is based on the following factorization lemma, which is also of interest from other points of view: An arbitrary operator $R: X^* \to X^{**}$ with the preceding property can be factorized into the form $R = A^*A$, where A is a bounded linear transformation of the space X^* into some auxiliary Hilbert space (in some sense the operator A plays the role of the square root of R).

Next, the transformations given by the covariance operators are investigated; and, in particular, different sufficient conditions are considered under which $RX^* \subset X$ (in the sense of the natural embedding of X into X^{**}).

Further, we look at the problem of the characterization of covariance operators of Gaussian distributions. Using the factorization $R = A^*A$, we reduce the problem to that of characterizing bounded linear transformations $A: X^* \to H$, where H is an auxiliary Hilbert space. We define in two ways Hilbert–Schmidt linear transformations of a Banach space into a Hilbert space (and conversely, a Hilbert space into a Banach space). Correspondingly, we obtain two definitions for the operator $R: X^* \to X^{**}$ to be a *nuclear operator*, and we prove that the first condition of being a nuclear operator is necessary and the second is sufficient for R to belong to the class of Gaussian covariance operators. By making some strongly restrictive assumptions on X, we find that both definitions of a nuclear operator are identical, and hence we obtain the necessary and sufficient condition. The example of the space l_p ($p \neq 2$) shows immediately that the assumptions in the given definitions are not directly related to the existing general definition of a nuclear transformation of one Banach space into another.

Finally, we show that the conditions for an operator R to belong to the class of Gaussian covariance operators are of a topological nature: There exists a Hausdorff locally convex topology in X^*, which generalizes the topology induced in the second chapter for l_q, and the necessary and sufficient condition is the continuity of the quadratic functional $(Rf)(f)$ with respect to this topology.

Note that some of the results in the fourth chapter could be made more general by considering linear topological spaces instead of Banach spaces. However, the whole style of the book and its relatively elementary level would be lost by doing this, and it would be too high a price to pay for generalizations of some separate particular results.

I would like to thank Yu. V. Prohorov and V. V. Sazonov for our discussions on the results considered in this book, as well as on other problems. This has been an opportunity I have enjoyed for many years.

Probability Distributions on Linear Spaces

1

Characteristic Functionals of Probability Distributions on Spaces of Numerical Sequences

Introduction

Section 1.1 is a general introduction, which gives a description of the space R^N and its conjugate (dual). Note that not all the properties given here will be used later on. However, investigation of different types of infinite-dimensional spaces, in which the method of characteristic functionals is applied, should be considered not only for their particular interest but also as the possible steps along the way toward explaining the most general situations where these methods could be applied. From this point of view, it is desirable to underline those properties that help to apply the method of characteristic functionals in each such case. The importance of this method is connected with the theorems of Bochner and Lévy, which were first proved in the one-dimensional case.

In Section 1.2 it is shown that for a distribution on R^N both these theorems are true in the classical formulation. Then we go to Banach spaces $X = l_p$ $(1 \leqslant p < \infty)$, c_0, and l_∞. We consider X not as a measurable space itself but as a measurable subset of R^N, and we obtain conditions under which

a. The distribution on R^N will be concentrated on X.
b. A compact family of distributions on R^N, each of which is concentrated on X, will also be compact in X.

By this method, in Sections 1.3 and 1.4, we obtain different theorems of the type of Bochner and Lévy, which include as particular cases some of the results of Yu. V. Prohorov and V. V. Sazonov for Hilbert spaces (this is described more completely in Section 3.1).

The main results of the first chapter are published in references 8 and 9.

In connection with the results of this chapter we note the survey-type article by Yu. V. Prohorov [10], in which is given a quite adequate description (with a large bibliography) of results of the method of characteristic functionals of distributions on linear topological spaces.

1.1 The Space R^N: Basic Properties

1.1.1

Recall that R^N is defined as the linear space of all real numerical sequences with the natural group operation (addition) and multiplication by real numbers. We shall say that the element $0 \neq x \in R^N$ has length L, if $x_k = 0$ for $k > L$ and $x_L \neq 0$. If such an L does not exist, then the length of this element will be infinity. The length of the zero element will be zero.

We shall assign to the linear space R^N the following (Tikhonov's) topology, taking as a base of neighborhoods of the zero element the class of sets of the form

$$\mathcal{O}_{\epsilon,n}(0) = \bigcap_{k=1}^{n} \{x : |x_k| < \epsilon\}, \quad \epsilon > 0, \quad n \geqslant 1.$$

The continuity of linear operations may be verified easily. Thus we obtain a linear topological space, which will be denoted R^N. We note the basic properties of this space.

1. R^N has a countable base, and therefore it is separable.
2. Convergence in R^N is equivalent to coordinate convergence.
3. R^N is a Fréchet space (F-space), that is, it is locally convex, metrizable, and complete.

The proofs of (1) and (2) are simple. Local convexity is obvious. To show that the space is metrizable we can take the following function as the metric.

$$\rho(x, y) = \sum_{k=1}^{\infty} \alpha_k \frac{|x_k - y_k|}{1 + |x_k - y_k|},$$

where the α_k are positive numbers that form a convergent series. The completeness is obvious.

4. R^N is additionally a *Montel–Fréchet space*, that is, along with (3), each closed bounded[1] set in R^N is compact.[2]

---

[1]The set $M \subset R^N$ being *bounded* is taken to mean: For each neighborhood of zero $\mathcal{O}_{\epsilon,n}(0)$ there exists $\lambda > 0$ such that

$$\{\lambda x : x \in M\} \subset \mathcal{O}_{\epsilon,n}(0).$$

[2]Note that because of the countable base, compactness and sequential compactness are equivalent for sets in R^N.

PROOF. The conclusion is a consequence of the well-known Bolzano–Weierstrass theorem and the fact that compactness and boundedness of sets in R^N mean correspondingly coordinatewise compactness and coordinatewise boundedness.

Note here that for relative compactness (also for boundedness) of a set M in R^N it follows (see also reference 11, p. 346) that it is necessary and sufficient that

$$\text{for all } x \in M, \qquad |x_k| \leqslant A_k, A_k > 0 \quad (k = 1, 2, \dots).$$

From this it follows that the space R^N is not locally compact. □

Now we want to find a space conjugate to R^N. By a *conjugate space* we mean a linear space (without fixing a topology) of all continuous linear functionals defined on R^N, using the usual operations of addition and scalar multiplication by reals (real numbers). Denote by R_0^N the subset of the space R^N, which consists of all elements with finite lengths.

5. The conjugate to R^N is R_0^N.

PROOF. It is sufficient to show that an arbitrary continuous linear functional f on R^N has the form $f(x) = \sum_k f_k x_k$, where the f_k are the coordinates of a fixed element $f \in R_0^N$ and are uniquely determined by the functional f. Obviously, x can be represented in the form

$$x = \sum_{k=1}^{n} x_k e^{(k)} + r^{(n)}x,$$

where $e^{(k)} = \{e_i^{(k)}\}$, with $e_i^{(k)} = 0$ for $i \neq k$ and $e_k^{(k)} = 1$, and the first n coordinates of the element $r^{(n)}x$ are equal to zero. Hence $r^{(n)}x \to 0$ as $n \to \infty$ for each element $x \in R^N$. Now, using the additivity and continuity of the functional f, and denoting the real number $f(e^{(k)})$ by f_k, we obtain from the previous equality the representation $f(x) = \sum_{k=1}^{\infty} f_k x_k$. We need to show that the element $f = \{f_k\}$ has a finite length. If this is not the case, then there exists an infinite subsequence $e^{(k_i)}$ such that $|f_{k_i}| > 0$. Then, $g^{(k_i)} = e^{(k_i)}/f_{k_i} \to 0$ for $i \to \infty$, whereas $f(g^{(k_i)}) = 1$ for all i, which contradicts the continuity. □

1.1.2

In the linear space R_0^N, which is a conjugate to a Fréchet space, one can introduce ([12], p. 74, although in this case it can be done explicitly) a locally convex topology I_c (called the *topology of uniform convergence on compact sets*) by defining as a base of neighborhoods of zero the family of sets of the form

$$U_{\epsilon,K}(0) = \left\{ f : \sup_{x \in K} |f(x)| < \epsilon \right\},$$

where ϵ is an arbitrary positive number and K is an arbitrary compact set in R^N.

1. The topology I_c can also be obtained from another base of neighborhoods of zero by taking sets

$$V_{\{\epsilon_k\}}(0) = \sum_{k=1}^{\infty} \{ f : |f_k| < \epsilon_k \},$$

where $\{\epsilon_k\}$ is an arbitrary sequence of positive numbers that converges to zero.

As a matter of fact, it is easy to see that $V_{\{\epsilon_k\}}(0) \subset U_{\epsilon,K}(0)$, if, for instance, $\epsilon_k = (\epsilon / k^2) A'_k$, where $A'_k = \max(1, A_k)$ and A_k $(k = 1, 2, \ldots)$ are numbers determined by the compact set K (see property (4) in Subsection 1.1.1. Conversely, $U_{\epsilon,K}(0) \subset V_{\{\epsilon_k\}}(0)$, if $\epsilon / A_k < \epsilon_k$ and K is the product of closed intervals $[0, A_k]$, because it is clear that in this case

$$\sup_{x \in K} |f(x)| \geqslant \sup_k |f_k| A_k.$$

2. The convergence of a sequence of elements $f^{(n)} \in R_0^N$ in the I_c-topology is equivalent to satisfying the following two conditions.
   a. Coordinatewise convergence.
   b. Boundedness of the lengths of the elements $f^{(n)}$.

Suppose $f^{(n)} \to 0$, because it is surely sufficient to consider only convergence to the zero element. The coordinatewise convergence is obvious. If the sequence of the lengths is not bounded, then there exists infinite subsequences of indices $\{n_j\}$ and $\{k_j\}$ such that $|f_{k_j}^{(n_j)}| > 0$. But then none of the elements $f^{(n_j)}$ will be in the neighborhood of zero $V_{\{\epsilon_j\}}(0)$ if $\epsilon_j \leqslant |f_{k_j}^{(n_j)}|$, $j = 1, 2, \ldots$, which contradicts the assumption that $f^{(n)} \to 0$. The converse is similarly easy to prove.

3. This convergence reduces to convergence in the so-called *weakest topology*, that is, convergence of $\{f^{(n)}\}$ as the weak convergence of linear functionals. The fact that convergence of $\{f^{(n)}(x)\}$ for all $x \in R^N$ follows from the convergence of $f^{(n)}$ in the I_c-topology is a simple consequence of being able to take the limit under the finite sum. The converse is nontrivial; that is, weak convergence implies convergence in the I_c-topology. For simplicity we consider only convergence to zero. The coordinatewise convergence follows from the convergence along the elements $e^{(k)}$. We shall show that the lengths are bounded. Without loss of generality, we assume the opposite—that the sequence of lengths $\{L_k\}$ corresponding to $\{f^{(k)}\}$ is strictly increasing. Then, we have $|f_{L_k}^{(k)}| > 0$, $k = 1, 2, \ldots$ and $L_k = +\infty$. But $f^{(k)}(x) \to 0$ for all x.

We now construct an element $a \in R^N$ on which the indicated sequence of functionals will not converge to zero, resulting in a contradiction. We shall determine the element a by means of the following relations (their solvability is obvious).

$$
\begin{aligned}
&a_k = 0 \quad \text{for} \quad k \neq L_j \quad (j = 1, 2, \ldots) \\
&a_{L_1} f_{L_1}^{(1)} = 1, \\
&a_{L_1} f_{L_1}^{(2)} + a_{L_2} f_{L_2}^{(2)} = 1 \\
&\vdots \quad \vdots \quad \vdots \quad \vdots \quad \vdots \qquad\qquad \vdots \quad \vdots \\
&a_{L_1} f_{L_1}^{(n)} + a_{L_2} f_{L_2}^{(n)} + \cdots + a_{L_n} f_{L_n}^{(n)} = 1,
\end{aligned}
$$

It is easy to see that $f^{(n)}(a) = 1$ for each n, and thus $f^{(n)}$ does not converge to zero along this element.

4. The convergence in the I_c-topology implies coordinatewise convergence as previously mentioned. The converse is not true (the simplest example is: $f_k^{(n)} = 1/n$ for $k \leqslant n$, $f_k^{(n)} = 0$ for $k > n$). This shows that the I_c-topology in R_0^N is stronger than the Tikhonov topology (induced from R^N).
5. The linear space R_0^N with the I_c-topology is a linear topological space. This space as a strong conjugate[3] to a Montel space is also a Montel space ([12], p. 90). The fact that in R_0^N boundedness is equivalent to compactness can be verified explicitly, noting that each compact (bounded) set in R_0^N consists of elements that have their lengths bounded by a common L and that the first L coordinates determine a compact (and respectively bounded) set in the Euclidean L-dimensional space. From this it follows, in particular, that the linear topological space (R_0^N, I_c) is not locally compact.
6. The conjugate to R_0^N is R^N. The topology of uniform convergence on I_c-compact sets in R_0^N (i.e., the topology $I_c = I_b$ in R^N) is identical with the original topology.

The proof is very simple, hence we omit it.

1.1.3

We now need to introduce the class of measurable sets in the space R^N. Sets in R^N will be called *measurable* if they belong to the minimal σ-algebra which contains all *Borel cylinders*, that is, sets of the form

$$\{x : x \in R^N, (x_1, x_2, \ldots, x_n) \in B\},$$

---

[3] By *strong conjugate* we mean the conjugate space with the *topology* I_b *of uniform convergence on bounded sets*. But it is obvious that $I_b = I_c$ in the case when the original space is a Montel space.

where n is any arbitrary natural number and B is an arbitrary Borel set in the Euclidean space R^n.

In the corresponding sense we shall also understand measurability of functionals determined in R^N, that is by taking preimages of all one-dimensional Borel sets to be measurable sets in R^N. All continuous linear functionals are measurable with respect to this σ-algebra, and this σ-algebra is minimal among those for which all of these functionals are measurable.

Moreover, all continuous functionals are measurable. This fact follows from the measurability of all Borel sets in the topological space R^N. The last statement follows from separability, that is, each open set is at most a countable union of neighborhoods, each of which is measurable. Therefore all open sets, and thus also all Borel sets, are measurable.

1.2 Probability Distributions on the Space R^N

1.2.1

A *probability distribution*, or, simply, a *distribution*, means a finite nonnegative countably additive measure, defined on a σ-algebra of measurable sets and normed so that the measure of the whole space R^N is one. Note that the limitation created by the normalization is not essential but is introduced only for simplicity.

In accordance with comments made in the introduction, the *characteristic functional* of the distribution μ is a functional in the dual space R_0^N, defined by the formula

$$\chi(f) = \chi(f;\ \mu) = \int_{R^N} \exp(if(x))\mu(dx).$$

REMARK. For characteristic functionals $\chi(f)$, the (uniform) continuity on the whole space R_0^N is assured by the continuity of $\chi(f)$ (or even only its real part) at zero, and therefore the equicontinuity of the family of characteristic functionals follows from the equicontinuity of this family at zero. This follows from the inequality

$$|\chi(f+h) - \chi(f)|^2 \leqslant 2|1 - Re\chi(h)|,$$

the proof of which does not differ from the proof in the finite-dimensional case ([13], p. 195).

The following theorem is analogous to Bochner's theorem and can easily be proved by applying Kolmogorov's theorem on extending a consistent system of finite-dimensional distributions to a measure.

1.2.2

Theorem. *In order for the functional* $\chi(f)$, $f \in R_0^N$ *to be the characteristic functional of some distribution on* R^N, *it is necessary and sufficient that the following conditions be satisfied.*

a. χ *is positive definite; that is, for each natural number* n, *for all collections of complex numbers* $a_1, a_2, \ldots, a_n$ *and elements* $f^{(1)}$, $f^{(2)}, \ldots, f^{(n)}$ *from* R_0^N, *the following inequality is satisfied.*

$$\sum_{i,j=1}^{n} a_i \bar{a}_j \chi(f^{(i)} - f^{(j)}) \geqslant 0,$$

b. $\chi(0) = 1$,
c. $\chi(f)$ *is continuous at* $f = 0$ *in the* I_c*-topology.*

If these conditions are satisfied, then the corresponding distribution is determined uniquely.

PROOF. *Sufficiency.* The continuity of $\chi(f)$ at zero in the I_c-topology implies the continuity $\chi(f)$ at zero on each finite-dimensional subspace, that is, the continuity of the function of n real variables $\chi(f_1, f_2, \ldots, f_n, 0, 0, \ldots) = \chi_n(f_1, \ldots, f_n)$ at zero for each fixed n. It is clear that χ_n is positive definite and that $\chi_n(0) = 1$. Therefore from the known finite-dimensional Bochner theorem it follows that there exists a unique distribution $\mu^{(1,2,\ldots,n)}$ in the n-dimensional Euclidean space, such that $\chi(\cdot; \mu^{(1,2,\ldots,n)}) = \chi_n(\cdot)$. It can easily be verified that the usual consistency conditions for the system of distributions $\{\mu^{(1,2,\ldots,n)}, n = 1, 2, \ldots\}$ are satisfied. Also, the proof of sufficiency can be completed by applying Kolmogorov's theorem on extending a consistent system of finite-dimensional distributions to a distribution on the σ-algebra of Borel cylinders, because this σ-algebra in R^N is identical (as we noted earlier) with the Borel σ-algebra.

REMARK. A system of finite-dimensional distributions $\mu^{(1,2,\ldots,n)}$ is called a *weak distribution* on R^N. Thus the sufficiency condition of this theorem states that each weak distribution in R^N can be extended uniquely to a Borel distribution (which is compact—to be discussed later).

We now return to the proof of the theorem.

Necessity. Condition (a) reduces to the corresponding condition for the finite-dimensional case (because of the continuity of $\chi(f)$ on each finite-dimensional subspace) and can be verified explicitly. Condition (b) is obvious. We now prove condition (c). Let $\epsilon > 0$ and consider a compact set K_ϵ in R^N such that $u(K_\epsilon) \geqslant 1 - \epsilon$. The existence of such a compact set for each $\epsilon > 0$ follows from the separability, metrizability, and completeness of the space R^N ([14], p. 40), which means, in the existing terminology, that each Borel distribution on R^N is a compact distribution. For $f \in U_{\epsilon, K_\epsilon}$ we have

$$1 - Re\chi(f) = \int_{K_\epsilon} [1 - \cos f(x)] \, \mu(dx) + \int_{R^N \setminus K_\epsilon} [1 - \cos f(x)] \, \mu(dx)$$

$$\leqslant \frac{1}{2} \sup_{x \in K_\epsilon} |f(x)|^2 + 2\epsilon \leqslant \epsilon^2/2 + 2\epsilon,$$

which shows the continuity of χ in the I_c-topology. □

1.2.3

REMARK. The characteristic functional is generally not continuous in the Tikhonov topology (induced from R^N). This is shown in the following example. Consider the functional

$$\chi(f) = \exp\left\{ -\frac{1}{2} \sum f_k^2 \right\}, \qquad f \in R_0^N.$$

This functional satisfies all the conditions of the preceding theorem and therefore is the characteristic functional of some distribution on R^N. The sequence of elements $f^{(n)}$, where

$$f_k^{(n)} = 1\sqrt{n} \quad \text{for } k \leqslant n, \qquad f_k^{(n)} = 0 \quad \text{for } k > n,$$

converges to zero in the Tikhonov topology (in other words, on each finite-dimensional subspace) as $n \to \infty$. However,

$$\chi(f^{(n)}) = \exp\left\{ -\frac{1}{2} \right\}$$

for all n, and therefore $\chi(f^{(n)})$ does not approach one as $f^{(n)}$ approaches zero.

1.2.4

In addition to Bochner's theorem there are other important theorems showing the utility of the method of characteristic functionals, mainly theorems that give conditions of weak convergence in terms of corresponding characteristic functionals. Before we go to these problems for the case of R^N, let us recall the definition of *weak convergence of distributions*.

Let X be an arbitrary separable metric (or metrizable topological) space.

The sequence of distributions $\{\mu_n\}$ in X is called *weakly convergent* to a distribution μ, if for each continuous and bounded functional F in X the following relation is satisfied.

$$\lim_{n\to\infty}\int_X F(x)\mu_n(dx)=\int_X F(x)\mu(dx).$$

Yu. V. Prohorov [4] has shown that weak convergence is equivalent to the convergence in the sense of some metric, often called the *Lévy–Prohorov metric*.

1.2.5

Later we shall use Prohorov's criterion [4] of weak relative compactness (in the sense of weak convergence) of families of distributions. Let us formulate this criterion.

For the weak relative compactness of the family $\{\mu_\alpha\}$ of distributions on a complete separable metric space, it is necessary and sufficient that for each $\epsilon>0$ there exists a compact set K_ϵ such that $\mu_\alpha(K_\epsilon)\geqslant 1-\epsilon$ for all α.

According to properties (1) and (3) from Subsection 1.1.1, it can easily be seen that this criterion is applicable to the case of distributions on R^N.

1.2.6

Theorem. *For the weak relative compactness of the family $\{\mu_\alpha\}$ of distributions on R^N, it is necessary and sufficient that the family of corresponding characteristic functionals $\{\chi_\alpha(f)=\chi(f;\mu_\alpha)\}$ be equicontinuous*[4] *in the I_c-topology.*

PROOF. *Sufficiency.* Let $\epsilon>0$ be arbitrary. Then, we can find a compact set K_ϵ having the following property: $\mu_\alpha(K_\epsilon)\geqslant 1-\epsilon$ for all α. As before, let $e^{(k)}$ be the point in R_0^N with coordinates $e_i^{(k)}=0$ for $i\neq k$ and $e_k^{(k)}=1$ $(k=1,2,\ldots)$. It is obvious that $\chi(te^{(k)};\mu_\alpha)$, as a function of t, $-\infty<t<+\infty$, is the characteristic function of a one-dimensional distribution $\mu_\alpha^{(k)}$, given by

$$\mu_\alpha^{(k)}(E)=\mu_\alpha\{x:x_k\in E\},$$

where E is an arbitrary Borel set of the real line.

From the assumptions of the theorem, the equicontinuity of the family $\{\chi(te^{(k)};\mu_\alpha)\}$ follows for an arbitrary fixed k. From this, and from the one-dimensional variant of the theorem, we obtain the weak conditional compactness of the family of one-dimensional distributions $\{\mu_\alpha^{(k)}\}$ for each

[4] See remark in Subsection 1.2.1.

k. Therefore, using Prohorov's criterion once more, we can say that for each $\epsilon_k > 0$ there exists a compact set $(K_{\epsilon_k}^{(k)})$ in R^1, such that $\mu_\alpha^{(k)}(K_{\epsilon_k}^{(k)}) \geqslant 1 - \epsilon_k$.

Let us take $\epsilon_k > 0$ $(k = 1, 2, \ldots)$ for a fixed ϵ so that $\sum_{k=1}^{\infty} \epsilon_k \leqslant \epsilon$. The set

$$\bigcap_{k=1}^{\infty} \left\{x : x_k \in K_{\epsilon_k}^{(k)}\right\},$$

being the product of one-dimensional compact sets, is compact in R^N according to Tikhonov's theorem. Denote this set by K_ϵ, which depends on ϵ. Then, we have

$$\begin{aligned}
\mu_\alpha(R^N \setminus K_\epsilon) &= \mu_\alpha\left(\bigcup_{k=1}^{\infty} \left\{x : x_k \in R^1 \setminus K_{\epsilon_k}^{(k)}\right\}\right) \\
&= \lim_{n\to\infty} \mu_\alpha\left(\bigcup_{k=1}^{n} \left\{x : x_k \in R^1 \setminus K_{\epsilon_k}^{(k)}\right\}\right) \\
&\leqslant \lim_{n\to\infty} \sum_{k=1}^{n} \mu_\alpha^{(k)}\left(R^1 \setminus K_{\epsilon_k}^{(k)}\right) \leqslant \lim_{n\to\infty} \sum_{k=1} \epsilon_k < \epsilon.
\end{aligned}$$

This ends the proof of sufficiency. □

The necessity can be shown similarly as in the theorem of Subsection 1.2.2. One should note that according to Prohorov's criterion, the compact set K_ϵ may be taken the same as for the whole family of distributions.

1.2.7

REMARK. The following fact proved in the sufficiency part of the theorem of Subsection 1.2.6 can be formulated as a separate statement: *the weak conditional compactness for all k of the family of one-dimensional distributions* $\{\mu_\alpha^k\}$ *implies the weak conditional compactness of the family* $\{\mu_\alpha\}$ *of distributions on* R^N.

From this remark and from the remark to the theorem of Subsection 1.2.2 (theorem of Subsection 1.2.9, which will be given later, can be used instead) it is easy to obtain the following theorem, which is the analogue of Lévy's continuity theorem, a result well known in the finite-dimensional case.

1.2.8

Theorem. *If the sequence* $\{\chi_n(f) = \chi(f; \mu_n)\}$ *of characteristic functionals is pointwise convergent to the limit* χ, *and* χ *is continuous at zero in the* I_c*-topology, then* χ *is the characteristic functional of some distribution* μ

and the sequence of distributions $\{\mu_n\}$ *is weakly convergent to the distribution* μ.

1.2.9

Theorem. *If the sequence of distributions is weakly compact and the sequence* $\{\chi(f;\mu_n)\}$ *is pointwise convergent to the limiting functional* χ, *then* $\{\mu_n\}$ *is weakly convergent to some distribution* μ *and* $\chi(f)=\chi(f;\mu)$.

PROOF. The proof of this theorem is simple and proceeds the same way in all cases: If $\{\mu_n\}$ is not convergent, then one can choose two subsequences convergent to different limits. But then from the sequence $\{\chi(f;\mu_n)\}$ one could also choose two subsequences convergent to different limits (according to the last statement of Subsection 1.2.2.). This contradicts the convergence of the sequence $\{\chi(f;\mu_n)\}$. The equality $\chi(f)=\chi(f;\mu)$ is obvious from the definition of weak convergence of distributions, because the functional $e^{if(x)}$ is bounded and continuous in R^N for each $f\in R_0^N$. □

1.2.10

At the end of this section we shall prove an inequality that connects the measure of the complement of a cylinder set, of which the base is an n-dimensional ball in the l_p-metric, $1\leqslant p\leqslant 2$, with a corresponding characteristic functional. This type of inequality can be quite useful for different purposes. For the case $p=2$ this inequality is basically identical to the one that was noticed by A. N. Kolmogorov in this work of Yu. V. Porhorov [4]. Kolomogorov [7] found this inequality as an independent result, which allowed him to combine the results of V. V. Sazonov and R. A. Minlos, as mentioned in the Introduction.

A similar inequality is given by L. Schwartz [15] by using Prohorov's method. It is well known that $e^{-|t|^p}$, $1\leqslant p\leqslant 2$, is the characteristic function of an absolutely continuous (with respect to Lebesgue measure) distribution on the real line. Let θ_p be the probability density function of this distribution.

Theorem. *For each distribution* μ *with the characteristic functional* χ, *for each natural number* n, *numbers* $r>0$ *and* $1\leqslant p\leqslant 2$ *the following inequality holds.*

$$\mu\left\{x:\sum_{k=1}^{n}|x_k|^p\geqslant r^p\right\}$$

$$\leqslant C\int\left[1-\mathrm{Re}\,\chi(f_1/r,\ldots,f_n/r,0,0,\ldots)\right]\prod_{k=1}^{n}\theta_p(f_k)\,df_k,$$

where C *is an absolute constant* $(C=e/(e-1))$.

PROOF. The proof is similar to the one for the case $p = 2$, given by V. V. Sazonov [16]. Let E denote the set on the left-hand side of the inequality. Now, in the integral over E replace unity by the function

$$C\left[1 - \exp\left(-\sum_{i=1}^{n} |x_j/r|^p\right)\right],$$

and extend integration to the whole space. We then obtain the inequality

$$\mu(E) \leqslant C\int\left[1 - \prod_{j=1}^{n} e^{-|x_j/r|^p}\right]\mu(dx)$$

$$= C\int\left[1 - \prod_{j=1}^{n}\int e^{if_j x_j/r}\theta_p(f_j)\,df_j\right]\mu(dx)$$

$$= C\int\left[1 - \int\exp\left(i\sum_{j=1}^{n}\frac{f_j x_j}{r}\right)\mu(dx)\right]\prod_{j=1}^{n}\theta_p(f_j)\,df_j,$$

which is equivalent to the previously stated one. □

1.3 Probability Distributions on Subspaces of the Space R^N

1.3.1

First of all we shall state the well-known notions and generally used conventions for different subspaces of the linear space R^N and some properties that will be used later.

Let $1 \leqslant p < \infty$ and

$$l_p = \left\{x : \sum_{j=1}^{\infty} |x_j|^p < +\infty\right\}.$$

This linear subspace of R^N is a Banach space if the norm is defined by

$$\|x\| = \|x\|_{l_p} = \left(\sum_{j=1}^{\infty} |x_j|^p\right)^{1/p}.$$

The space conjugate to l_p is the space l_q, where q is the conjugate parameter such that $1/p + 1/q = 1$. The space conjugate to l_1 is the subspace l_∞, which is defined by

$$l_\infty = \{x : \sup|x_j| < +\infty\}$$

and is a Banach space with the norm

$$\|x\| = \|x\|_{l_\infty} = \sup_j |x_j|.$$

The space l_∞ is often denoted by m.

Further, let

$$c_0 = \left\{ x : \lim_{j\to\infty} x_j = 0 \right\}.$$

The space c_0 is a subspace of the space l_∞ and with the norm induced from l_∞ is also a Banach space. The conjugate to c_0 is l_1.

The spaces c_0, l_1, l_∞ are nonreflexive, the spaces l_p $(1 < p < \infty)$ are reflexive. The space l_∞ is not separable, whereas c_0 and l_p $(1 \leqslant p < \infty)$ are separable. We have $l_{p'} \subset l_p$ if $p' < p$, and $l_p \subset c_0$ for each p, $1 \leqslant p < \infty$.

In addition, it will be assumed everywhere that p and q are conjugate parameters, that is, $1/p + 1/q = 1$ and for $p = 1$, $q = \infty$.

1.3.2

The space l_2 is a Hilbert space, and as a Hilbert space it is universal; that is, for each real separable Hilbert space H there exists an isometric isomorphism (a one-to-one correspondence that preserves linear operations and the inner product) between H and l_2. We shall use this fact (which is the Riesz–Fischer theorem) later.

1.3.3

Borel cylinder sets of some subspace X of the space R^N can be defined as a *trace* in X of the Borel cylinder sets in R^N, that is, the collection of sets each of which is the intersection of X with some Borel cylinder set in R^N. In other words, the set $E \subset X$ is a *cylinder set* if there exists a natural number $n = n(E)$, such that the fact that an element $x \in X$ belongs to E can be expressed by conditions put on the first $n(E)$ coordinates of the elements in X. If, in addition, these conditions determine a Borel set in the Euclidean space $R^{n(E)}$, then E is called a *Borel cylinder set*. This definition is consistent with the definition of the Borel cylinder set in linear topological spaces. This is not difficult to show, and therefore we omit the proof.[5]

The class of all Borel cylinder sets is a *Boolean algebra*. We shall always

---

[5]For the case of $X = l_p$ $(1 \leqslant p < \infty)$ or $X = c_0$ we obtain the σ-algebra generated by the whole dual space, whereas for $X = l_\infty$ the intersection of X with the σ-algebra of Borel sets in R^N gives the σ-algebra generated by the linear functionals from $l_1 \subset l_\infty^*$.

consider the σ-algebra generated by this algebra (i.e., the smallest one that contains it). In particular, the term *distribution* (*probability distribution*) on X will always mean a nonnegative countably additive normed measure determined on this σ-algebra. Sets that are members of this σ-algebra will be called *absolutely measurable* or, simply, *measurable*. If a particular distribution μ is considered, then μ-measurability means belonging to the σ-algebra that is completed by all possible subsets of all measurable sets with μ-measure zero. Similarly, we have the measurability and μ-measurability of functionals determined in X. The need of considering the σ-algebra generated by Borel cylinder sets is explained by the existence of a sufficiently rich class of measurable functionals. Thus let us formulate the facts along these lines, which we shall often use without further comment.

1. In the spaces l_p $(1 \leqslant p < \infty)$ and c_0, each continuous linear functional is measurable. In the space l_∞ all continuous linear functionals from the space $l_1 \subset l_\infty^*$ are measurable. Note further, that σ-algebras of Borel cylinder sets in these spaces are minimal among those for which the aforementioned linear functionals are measurable.

The proof of measurability follows easily from the fact that the pointwise limit of measurable functionals is measurable. The property of minimality is a simple consequence of the measurability of the coordinate linear functionals (i.e., functionals $e^{(k)}$, $k = 1, 2, \ldots$).

2. In the spaces l_p $(1 \leqslant p \leqslant \infty)$ and c_0 the norm of an element is a measurable functional. This follows from the relations:

$$\{x : \|x - a\|_{l_p} \leqslant r\} = \bigcap_{n=1}^{\infty} \left\{x : \sum_{j=1}^{n} |x_j - a_j|^p \leqslant r^p\right\} \quad 1 \leqslant p < \infty,$$

$$\{x : \|x - a\|_{l_\infty, c_0} \leqslant r\} = \bigcap_{n=1}^{\infty} \{x : \max_{1 \leqslant j \leqslant n} |x_j - a_j| \leqslant r\},$$

in which $r > 0$ and a is an arbitrary element (in particular, one can take $a = 0$).

3. From these relations it follows that all balls in these spaces are measurable. Hence by taking into account the separability of the spaces l_p $(1 \leqslant p < \infty)$ and c_0, we obtain the fact that the class of measurable sets in these spaces is identical with the class of Borel sets in them if these spaces are treated as Banach spaces.

Thus in the spaces l_p $(1 \leqslant p < \infty)$ and c_0, each continuous functional is measurable. In the space l_∞ this is not true.[6]

---

[6] In this case the smallest σ-algebra, with respect to which all continuous functionals are measurable (i.e., Baire's σ-algebra), is larger than the σ-algebra generated by Borel cylinder sets and smaller than the σ-algebra generated by open sets (the Borel σ-algebra). For conditions of uniqueness of the extension of a distribution to the Baire or Borel σ-algebra, see Yu. V. Prohorov [10].

4. The set $E \subset l_p$ $(1 \leqslant p \leqslant \infty)$ or $E \subset c_0$ is measurable in the corresponding spaces if and only if it is measurable as a set in R^N.

To prove this let X denote any of the spaces under consideration. It is easy to show, in the standard way, that the class of measurable sets of the space X is identical to the intersection of X with the class of measurable sets in R^N. Therefore X is a measurable set in R^N; and if $E \subset X$ is measurable in X, then $E = E' \cap X$, where E' is measurable in R^N. Then E is measurable in R^N as the intersection of two measurable sets in R^N. Conversely, if $E \subset X$ is measurable in R^N, then $E = E \cap X$ and therefore is also measurable in X.

From this property it follows that the embeddings $l_{p'} \to l_p$ $(p' < p)$, $l_p \to c_0$, $c_0 \to l_\infty$ are measurable transformations; that is, for example, a measurable set in $l_{p'}$ considered as a set in l_p ($p' < p$; and therefore $l_{p'} \subset l_p$) is a measurable set in l_p. Also, it follows from this property that if μ is a distribution on R^N, which is concentrated on X (i.e., $\mu(X) = 1$), then μ is also a distribution on X. Conversely, each distribution μ on X can be treated as a restriction of some distribution $\tilde{\mu}$ on R^N, taking $\tilde{\mu}(E) = \mu(E \cap X)$ for each measurable set E.

1.3.4

Now we shall give simple sufficient conditions under which a distribution μ on R^N is concentrated on some of its subspaces.

Theorem.

1. *If* $\sum_{k=1}^{\infty} \int_{R^1} |u|^p \mu^{(k)}(du) < +\infty, \quad 1 \leqslant p < \infty,$ *then* $\mu(l_p) = 1$.
2. *If* $\lim_{n\to\infty} \int_{R^n} \max_{1 \leqslant j \leqslant n} \{|u_j|\}\, \mu^{(1,2,\ldots,n)}(du) < +\infty,$ *then* $\mu(l_\infty) = 1$.
3. *If* $\lim_{n\to\infty} \lim_{k\to\infty} \int_{R^k} \max_{n < j \leqslant n+k} \{|u_j|\}\, \mu^{(n+1,\ldots,n+k)}(du) = 0,$ *then* $\mu(c_0) = 1$.

In this theorem μ with upper indices means the corresponding finite-dimensional distribution (projection of the distribution μ). Thus

$$\mu^{(i,i+1,\ldots,i+s)}(E) = \mu\{x : x \in R^N, (x_i, x_{i+1}, \ldots, x_{i+s}) \in E\},$$
$$(i \geqslant 1, s \geqslant 0),$$

for Borel sets E in the space R^{s+1}.

PROOF. For the proof of the first two conclusions we use the theorem on the limit of an increasing sequence of integrable functions, noting that an integral with respect to the measure μ of a functional in R^N, which depends only on a finite number of coordinates, can be reduced to the integral on the corresponding projection of the measure μ. □

For the proof of the third conclusion write c_0 as

$$c_0 = \bigcap_{m=1}^{\infty} \bigcup_{n=1}^{\infty} \bigcap_{k=1}^{\infty} \left\{ x : \max_{n<j\leqslant n+k} \{|x_j|\} < 1/m \right\}$$

and use Chebyshev's inequality, noting once more that integration of a measure on a part of the coordinates gives the projection on the remaining part. □

1.3.5

These conclusions can be expressed in terms of weak distributions. A *weak distribution* on a subspace $X \subset R^N$ is a nonnegative finitely additive set function, determined on an algebra of Borel cylinder sets from the space X and countably additive on the subalgebra, consisting of Borel cylinder sets determined by a fixed set of coordinates. Unlike the case of the space R^N, a weak distribtuion on X cannot always be extended to a probability distribution because the property of upper continuity on the empty set, which guarantees the countable additivity, is not always true (Borel cylinder sets $E_1 \supset E_2 \supset \cdots \supset E_n \supset \cdots$ from X can obviously be considered as Borel cylinder sets in R^N, but from the fact that $\bigcap E_i$ is empty in X it does not follow that this intersection is empty in R^N.)

But a weak distribution on X is also a weak distribution in R^N and therefore, according to the theorem of Subsection 1.2.2, it can be extended (in a unique way) to a probability distribution on R^N, and applying the conclusions from the previous section we have the following theorem.

Theorem. *A weak distribution on* l_p $(1 \leqslant p < \infty)$, l_∞, *or* c_0 *can be uniquely extended to a probability distribution if condition* (1), (2), *or* (3) *from the theorem of Subsection* 1.3.4 *is satisfied.*

1.3.6

In this section we shall show that if the family $\{\mu_\alpha\}$ of distributions on R^N satisfies conditions (1) or (2) of the theorem of Subsection 1.3.4 uniformly in α, then not only will each μ_α be concentrated on the corresponding subspace but (by some natural additional assumptions) the family will be conditionally compact in this subspace.

Theorem. *For the weak conditional compactness of the family* $\{\mu_\alpha\}$ *of distributions on* l_p $(1 \leqslant p < \infty)$ *or correspondingly on* c_0, *it is sufficient that the following conditions be satisfied.*[7]

---

[7] Satisfying these conditions means conditional compactness in l_1 of the set $\{m^{(\alpha)}\}$, where $m^{(\alpha)}$ is the pth absolute moment of the distribution μ_α.

a. $\sup_\alpha \sum_{k=1}^{\infty} \int_{-\infty}^{+\infty} |u|^p \mu_\alpha^{(k)}(du) \leqslant C$,
b. $\lim_{n\to\infty} \sup_\alpha \sum_{k=n}^{\infty} \int_{-\infty}^{\infty} |u|^p \mu_\alpha^{(k)}(du) = 0$,

or, correspondingly, the conditions:

a′. $\sup_\alpha \lim_{n\to\infty} \int_{R^n} \max_{j\leqslant n} \{|u_j|\}\, \mu_\alpha^{(1,2,\ldots,n)}(du) \leqslant C$,
b′. $\lim_{n\to\infty} \sup_\alpha \lim_{k\to\infty} \int_{R^n} \max_{n<j\leqslant n+k} \{|u_j|\}\, \mu_\alpha^{(n+1,\ldots,n+k)}(du) = 0$.

Here C is a positive constant, and the projections of the measure μ_α are determined the same way as in the theorem of Subsection 1.3.4.

PROOF. The proof is based on Prohorov's criterion, given in Section 1.2.5. We shall prove the first part concerning the spaces l_p. The second part may be proved similarly.

It is clear that for each $r>0$, and for each increasing sequence N_m of natural numbers, the set

$$K = \mathcal{O}_r \bigcap_{m=1}^{\infty} V_m,$$

where $\mathcal{O}_r$ is a ball in l_p, with radius r and center zero and where

$$V_m = \left\{ x : x \in l_p, \sum_{k=N_m}^{\infty} |x_k|^p \leqslant 1/m \right\},$$

is a compact set in l_p. Applying Chebyshev's inequality and the theorem on integrating a series of nonnegative functions term by term, we obtain

$$\mu_\alpha(l_p \backslash K) \leqslant Cr^{-p} + \sum_{m=1}^{\infty} m\psi(N_m),$$

where $\psi(n)$ means the expression that is under the limit in condition (b) of the theorem. Now, choosing for an arbitrary $\epsilon>0$ a sequence $N_m \to \infty$ so that

$$\sum^{\infty} m\psi(N_m) < \epsilon/2$$

(this can be done because of condition (b)) and taking $r \geqslant (2C)^{1/p}\epsilon^{-1/p}$, we obtain the condition

$$\mu_\alpha(l_p \backslash K) \leqslant \epsilon;$$

and by applying Prohorov's criterion, we complete the proof. □

1.4 Characteristic Functionals of Distributions on Subspaces of the Space R^N

1.4.1

First, let us recall a definition. Let X denote the space c_0 or one of the spaces l_p $(1 \leqslant p < \infty)$; let μ be a distribution on X. If X is c_0 or l_p for $1 \leqslant p < \infty$, then the *characteristic functional* is defined as a functional in the conjugate space X^* by the formula

$$\chi(f) = \int_X e^{if(x)}\mu(dx), \qquad f \in X^*. \tag{1.1}$$

Now, let X be l_1 or l_∞. Then $X = Y^*$, where Y is correspondingly c_0 or l_1, and χ is determined as a functional in Y, again by formula (1.1), in which $f \in Y \subset X^*$.

Of course, in the case $X = Y^*$ one can determine χ as a functional in the conjugate space $X^* = Y^{**}$, but this would mean the investigation of the general theory without taking into account the properties of the particular spaces; although it would give us a uniform definition, some technical complications would be created. As a matter of fact, this is not necessary, for both definitions give the same values of χ on $Y \subset Y^{**}$, and the knowledge of χ on this part is sufficient because the commonly used properties that make characteristic functionals useful are still satisfied. These properties as well as a more detailed investigation of these problems will be given in Chapter 4. We shall only note here that both definitions of the characteristic functional (as a functional in Y^{**} or only in Y) are the same for l_p for $1 < p < \infty$ (because of reflexivity), and for l_1 the values of χ on c_0 uniquely determine the values of χ on the whole of $c_0^{**} = l_\infty$ (see theorem of Subsection 4.1.6). Hence these two definitions of the characteristic functional also coincide for $X = l_1 = c_0^*$. In the case of c_0 the second definition does not make sense.

1.4.2

Theorem. *For a functional $\chi(f)$, $f \in l_q$ to be the characteristic functional of some distribution μ on l_p $(1 \leqslant p < \infty)$, such that*

$$\int_{l_p} \|x\|^p \mu(dx) < +\infty,$$

it is necessary and sufficient that the following conditions are satisfied.

a. *χ is positive definite.*
b. $\chi(0) = 1$.
c. *χ is continuous in the norm*[8] *in the space l_q.*

---

[8]The continuity at zero is sufficient (see Subsection 1.2.1, which is applicable also in this case).

d. $\sum_{k=1}^{\infty} \int_{-\infty}^{+\infty} |u|^p \mu^{(k)}(du) < +\infty$, *where* $\mu^{(k)}$ *is the distribution on the real line with the characteristic function* $\varphi_k(t) = \chi(te^{(k)})$.

If these conditions are satisfied, then the corresponding distribution is uniquely determined.

PROOF. *Sufficiency.* It is obvious that $l_q \supset R_0^N$, $c_0 \supset R_0^N$. Let χ_0 denote the restriction of χ to R_0^N. According to conditions (a), (b), and (c), the functional χ_0 satisfies all conditions of the theorem of Subsection 1.2.2. Therefore there exists a unique distribution μ on R^N with $\chi(f; \mu) = \chi(f)$, $f \in R_0^N$. According to condition (d) and to the theorem of Subsection 1.3.4, the distribution μ is concentrated on l_p. Now, it is enough to note that the equality $\chi(f) = \chi(f; \mu)$, which was established only for $f \in R_0^N$, remains true for all $f \in l_q$. This follows from the two facts: (1) R_0^N is everywhere dense in l_q and c_0; (2) The functionals $\chi(f)$ and $\chi(f; \mu)$ are both continuous in norm, the first according to the assumption, the second by the recent conclusion that $\mu(l_p) = 1$.

The necessity of condition (d) follows from the integrability of $\|x\|^p$ and can easily be obtained by using the known theorem about term-by-term integration of a series of nonnegative functions. The last conditions can be verified explicitly, as in the finite-dimensional case. □

1.4.3

Similarly, the following theorem can be proved.

Theorem. *For the functional* $\chi(f)$: $f \in l_1$ *to be the characteristic functional of some distribution* μ *on* l_∞ *for which*

$$\int_{l_\infty} \|x\| \, \mu \, dx < +\infty,$$

it is necessary and sufficient that the following conditions be satisfied.

a. χ *is positive definite.*
b. $\chi(0) = 1$.
c. χ *is continuous in the norm.*[9]
d. $\lim_{n\to\infty} \int_{R^n} \max\{|u_j|\} \, \mu^{(1,2,\ldots,n)}(du) < \infty$, *where* $\mu^{(1,2,\ldots,n)}$ *is a distribution on* R^N, *having the characteristic function*

$$f(t_1, t_2, \ldots, t_n) = \chi(t_1 e^{(1)} + t_2 e^{(2)} + \cdots + t_n e^{(n)}).$$

If these conditions are satisfied, then the corresponding distribution is uniquely determined.

We shall also note a result for the space c_0. Unlike l_p $(1 \leqslant p < \infty)$, the space c_0 as a subset of R^N, is not determined by the condition of finiteness

[9] See footnote 8.

of the functional that expresses the norm (the set $\{x : x \in R^N, \|x\|_{c_0} < +\infty\}$ is not c_0, but is l_∞). This fact implies that the theorem analogous to the two previous ones is not valid for c_0 in the necessity part; but the analogy is retained for the sufficiency part.

1.4.4

Let $\{\mu_\alpha\}$ be a family of distributions on c_0 or on l_p $(1 \leqslant p < \infty)$ and let $\{\chi_\alpha\}$ be the family of corresponding characteristic functionals. Repeating the similar considerations as in the proof of necessity in the theorem of Subsection 1.2.2, one can easily show that if $\{\mu_\alpha\}$ is weakly conditionally compact, then $\{\chi_\alpha\}$ is equicontinuous. In the case of the space R^N, as in the finite-dimensional case, the converse is also true (theorem of Subsection 1.2.2). But for subspaces of R^N this is not true: Equicontinuity in norm does not imply weak conditional compactness, and if one weakens the topology enough so that the converse becomes true, then the direct assertion becomes false. This fact was discovered by Yu. V. Prohorov and V. V. Sazonov [17], who showed that in the space l_2 there is no locally convex topology such that equicontinuity of the family $\{\chi_\alpha\}$ is equivalent to the weak conditional compactness of the family $\{\mu_\alpha\}$. Repeating the same considerations and using the form of Gaussian distributions given in the next chapter, one can obtain an analogous negative result for the spaces l_p $(1 \leqslant p < \infty)$ and c_0. However, considering distributions on these spaces as distributions on R^N, one can easily obtain a positive result; namely, a weakened analogue of the theorem of Subsection 1.2.2.

Theorem. *Let $\{\mu_\alpha\}$ be a family of distributions on c_0 or on l_p $(1 \leqslant p < \infty)$ and let $\chi_\alpha(f) = \chi(f; \mu_\alpha)$. If $\{\mu_\alpha\}$ is weakly conditionally compact, then $\{\chi_\alpha\}$ is equicontinuous (in norm). Conversely, if $\{\chi_\alpha\}$ is equicontinuous, then $\{\mu_\alpha\}$ is conditionally compact in the sense of convergence $\mu_n \to \mu$ determined by the condition*

$$\lim_{n\to\infty} \int F(x)\mu_n(dx) = \int F(x)\mu(dx) \tag{1.2}$$

for each bounded functional F (in c_0 or in l_p, respectively) continuous in the topology induced from R^N.[10]

PROOF. The proof is simple. One can use the theorem of Subsection 1.2.2 and note that the I_c-topology of the space R_0^N is stronger than the topology that is induced in R_0^N from l_1, c_0, and l_q, respectively. □

[10] Let us recall that weak convergence means that condition (1) is satisfied for each bounded and continuous functional. According to the fact that the topology induced from R^N is weaker than the metric topology, the convergence that is proved in the theorem is weaker than the weak convergence.

2

Gaussian Distributions on Spaces of Numerical Sequences

Introduction

In Section 2.1 we review some known properties of the Gaussian distributions that will be used later. Also, we pose the problem of finding the general form of characteristic functionals of all Gaussian distributions and solve the problem for the simple case of the space R^N. Using the method of embedding into R^N, we solve the problem in Section 2.2 for the case of the spaces l_p $(1 \leqslant p < \infty)$. The result obtained for $p = 2$ (theorem of Subsection 2.2.6) remains true for an arbitrary real separable Hilbert space H, and it is identical in this case with the result of E. Mourier. Embedding the space H into the scale of the spaces l_p makes the result for the space H more understandable and more natural (see Section 3.1).

Further, in this section we construct a *locally convex Hausdorff topology* $\mathfrak{T}_p$ in l_q $1 < q \leqslant \infty$ and prove the following theorem: A *Gaussian distribution on* R^N *is concentrated on* l_p *if and only if its characteristic functional is continuous in the* $\mathfrak{T}_p$*-topology*. This theorem shows, particularly, that $\mathfrak{T}_p$ is the natural generalization of the Sazonov's topology and is identical with it for $p = 2$. Another generalization of the Sazonov's theorem (this theorem will be proved in Section 3.1) has been given recently by L. Schwartz [18–19], who investigated the case $1 \leqslant p \leqslant 2$. The case $2 \leqslant p < \infty$ has been investigated by S. Kwapien [20].

Finally, in Section 2.2 we consider the connection between the existence of moments of a distribution and the rate of decrease of the diagonal elements of its covariance matrix. In particular, by using the theorem of Subsection 2.2.6, we obtain as a corollary a negative answer to a question posed in the work of R. Fortet and E. Mourier [21].

In Section 2.3 we extend our investigations of Gaussian distributions on l_p. We prove the central limit theorem (theorem of Subsection 2.3.2), which includes *Mourier's theorem* ($p = 2$) given in reference 3 and *Varadarajan's theorem* ($p = 1$), given in reference 22. We prove the integrability of any power of the norm with respect to an arbitrary Gaussian distribution and obtain an exponential estimate for the measure of the complement of any ball. In this way we derive the exponential integrability of the square of the norm. Applying the Fourier transform method, we investigate the Gaussian distribution corresponding to the solution of a stochastic heat conduction equation. This shows how effectively the basic theorem (Subsection 2.2.5) of this chapter may be used. Moreover, this example indicates that when applying embedding in R^N we should consider not only the image in R^N but also the form of the operator that gives the embedding.

Further, in Section 2.3 we define a nondegenerate distribution and give the characteristic property for nondegeneracy of the Gaussian distribution μ on l_p ($1 \leqslant p < \infty$): The μ-measure of an arbitrary ball with a nonzero radius is positive. Although this property of Gaussian distributions is very natural, it characterizes not only the nondegenerate Gaussian distributions in general but, rather, the space that carries the distributions. In Section 2.4 we prove that the nondegenerate Gaussian distributions do not have this property on the space l_∞. Moreover, in this section, sufficient conditions are given, under which the Gaussian distributions on R^N are concentrated on l_∞ or, correspondingly, on c_0, and necessity is proved by the additional assumption that the covariance matrix is diagonal.

The main results of this chapter were published in previously cited articles [8–9, 23–26]. In connection with the results of this chapter, we would also like to mention the article by R. Fortet [27] and the book by G. E. Shilov and Fan Dik Tin [28]. Among other articles on Gaussian distributions on linear spaces, we mention the article by A. M. Vershik [29].

2.1 The Definition and Basic Properties of Gaussian Distributions: Gaussian Distributions on the Space R^N

2.1.1

Let us start by recalling some general facts. A distribution on the finite-dimensional space R^n is called *Gaussian* if it is absolutely continuous (with respect to the ordinary Lebesgue measure) and the probability density function is

$$\Phi(x) = (2\pi)^{-n/2}(\det S)^{-1/2}\exp\left\{-\frac{1}{2}\left(S^{-1}(x-a),(x-a)\right)\right\}, \quad (2.1)$$

where $a \in R^n$ and S is a positive-definite matrix (S^{-1} denotes its inverse). The element a and the matrix $S = \|s_{ij}\|$ are the expectation and the covariance matrix of the Gaussian distribution, respectively, that is, they are given by the formulas

$$a_j = \int_{R^n} x_j \Phi(x)\, dx,$$

$$s_{ij} = \int_{R^n} (x_i - a_i)(x_j - a_j)\Phi(x)\, dx, \quad (i, j = 1, 2, \ldots, n).$$

Instead of (or together with) the notion of the Gaussian distribution, we can use the notion of a *Gaussian random vector*, which is a measurable[1] transformation $x = x(\omega)$ of some probability space $(\Omega, \mathfrak{B}(\Omega), P)$ into R^n and has a Gaussian distribution, such that for each Borel set E in R^n we have

$$P\left\{\omega : x(\omega) \in E\right\} = \int_E \Phi(x)\, dx.$$

The Gaussian distribution can also be defined by starting from the characteristic function given by

$$\chi(f) = \int_{R^n} e^{i(f,x)} \Phi(x)\, dx, \qquad f \in R^n. \tag{2.2}$$

Substituting (2.1) into (2.2) and integrating, we obtain the known formula for the characteristic function of the Gaussian distribution.

$$\chi(f) = \exp\left\{ i(f, a) - \frac{1}{2}(Sf, f) \right\}. \tag{2.3}$$

From this expression it can be seen that the assumption of positive definiteness of the matrix S can be weakened to nonnegative definiteness: $(Su, u) \geqslant 0$ for all $\mu \in R^n$. In the case of degeneracy ($\det S = 0$) the density that corresponds to the characteristic function (2.3) can be found as the limit as $\epsilon \to 0$ of the density $\Phi_\epsilon(x)$, which would differ from (2.1) only in that S^{-1} is replaced by $(S + \epsilon I)^{-1}$, where I is the identity matrix.

2.1.2

The preceding definition of the Gaussian distribution on a finite-dimensional space is not applicable for explicit generalization to the infinite-dimensional case. The definition through densities loses its meaning because there is no Lebesgue measure on the infinite-dimensional space; and the definition through the characteristic function cannot be generalized explicitly because the conditions on the infinite-dimensional covariance

[1] This is with respect to the Borel σ-algebra in R^n and the σ-algebra $\mathfrak{B}(\Omega)$ in Ω.

matrix depend on the space and this dependence is neither trivial nor clear beforehand. Determining the conditions that characterize covariance matrices in various infinite-dimensional spaces is the main problem in the investigation of Gaussian distributions. The problem for the case of different subspaces of the space R^N is solved in this chapter. The general case of Banach spaces is considered in the fourth chapter.

2.1.3

Let μ be the Gaussian distribution on R^n, let E_μ denote integration with respect to the measure μ, and let

$$\chi^{(f)}(t) = E_\mu e^{it(f,x)}, \qquad -\infty < t < \infty, \quad f \in R^n$$

be the characteristic function of the linear functional (f,x), considered as the usual (one-dimensional) random variable in (R^n, μ).

Using the definition of the characteristic function we obtain

$$\chi^{(f)}(t) = \chi(tf). \tag{2.4}$$

By using expression (2.3) we see that for each fixed f the random variable (f,x) has a Gaussian distribution[2] (on the real line) with parameters

$$E_\mu(f,x) = (f,a), \qquad E_\mu(f,x-a)^2 = (Sf,f).$$

Therefore if μ is a Gaussian distribution on R^n, then each linear continuous functional has a Gaussian distribution on the real line. The converse is also true: If each linear continuous functional in (R^n, μ) has a Gaussian distribution on the line, then μ is Gaussian. This follows easily from relation (2.4). Taking into account the form of the characteristic functions of Gaussian random variables and putting $t=1$ in (2.4), we obtain

$$\chi(f) = \exp\left(ia_f - \frac{1}{2}\sigma_f^2\right),$$

where

$$a_f = E_\mu(f,x), \qquad \sigma_f^2 = E_\mu(f,x-a)^2. \tag{2.5}$$

In order to obtain formula (2.3) for $\chi(f)$, one has only to note that a_f and σ_f^2, defined by (2.5), are, respectively, linear and quadratic functionals in f.

---

[2]To avoid mentioning each time that $f \neq 0$, it is convenient to treat the distribution concentrated at one point as a particular case of the Gaussian distribution (which corresponds to the variance being zero).

We note that for a random vector to be Gaussian it is not sufficient that the components be Gaussian. An example of this is given at the end of this section.

2.1.4

The characteristic property just given is usually taken as the basis for the definition of Gaussian distributions in general locally convex linear topological spaces. According to this definition, the distribution μ on R^N will be called *Gaussian* or *normal* (these terms are synonymous) if all the random variables $f(x)$, $f \in R_0^N$ determined in (R^N, μ), have Gaussian distributions on the real line.

Repeating the considerations of the previous subsection, one can see that this condition is equivalent to the natural condition that all finite-dimensional projections of the distribution μ are Gaussian (the definition of finite-dimensional distribution on R^N is given in Subsection 1.3.4). Or speaking in terms of random elements, one can see that the normality of the random sequence $\{x_k(\omega)\}$ in R^N means normality of all random vectors $(x_i(\omega), x_{i+1}(\omega), \ldots, x_{i+s}(\omega))$ in the corresponding finite-dimensional spaces.

This form of the definition of normality is also convenient for subspaces of R^N and can also be used without any change for the case: A distribution μ on some subspace $X \subset R^N$ is called *Gaussian* if all its finite-dimensional projections are Gaussian; or, in other words, a random sequence $\{x_k(\omega)\}$ with values in X is called a *Gaussian random element* in the space X if the joint distribution of an arbitrary number of elements of this sequence is Gaussian.

It is easy to show that this definition is also consistent with the definition of the Gaussian distribution by the set of continuous linear functionals given earlier. The fact that Gaussian distributions of all continuous linear functionals implies the Gaussian property in this sense is a simple conclusion of the obvious relation $R_0^N \subset X^*$ (we assume, of course, that the space X is infinite dimensional). The converse needs some additional argument when X is the conjugate of a nonreflexive space, but for our aims it is not necessary. In accordance with the contents of Subsections 1.3.3 and 1.4.1, we shall use only the fact that if μ is a Gaussian distribution on the space l_∞, then all the linear functionals $f \in l_1$ are Gaussian. If μ is a Gaussian distribution on the space c_0 or l_p for $1 \leqslant p < \infty$, then all linear continuous functionals have Gaussian distributions. This can be seen as true by noting that all such functionals can be approximated (at least in the sense of pointwise convergence) by finite length ($\in R_0^N$) functionals, having, by assumption, Gaussian distributions, and also that the limit (almost sure) of Gaussian random variables is again a Gaussian random variable.

2.1.5

Our next problem is a full description of all Gaussian distributions on the space R^N and its different subspaces. To do this we shall use the method of characteristic functionals.

The problem of the description of all Gaussian distributions in finite-dimensional spaces has the following known solution: The characteristic function is given by formula (2.3) in Subsection 2.1.1; and we obtain all Gaussian distributions on R^n if we take all possible $a \in R^n$ and all possible symmetric nonnegative-definite $n \times n$ matrices S.

In the infinite-dimensional case, as can be shown by simple considerations, this form of the characteristic functional should be retained. With the corresponding changes of the parameters, the problem, then, is finding the characteristic properties giving the admissible values of these parameters. Here we shall give the solution of this problem in the simple case of the whole space R^N, where the conditions of admissibility are reduced to the natural necessary conditions. The case of subspaces of the space R^N will be considered in subsequent sections.

Theorem. *The characteristic functional of an arbitrary Gaussian distribution on R^N has the form*

$$\chi(f) = \exp\left\{ i\sum_k f_k a_k - \frac{1}{2} \sum_{i,j} s_{ij} f_i f_j \right\}, \tag{2.6}$$

where $\{a_j\} \in R^N$ and the infinite matrix $\|s_{ij}\|$ is symmetric and nonnegative definite (the double sum in (2.6) is nonnegative for each $f \in R_0^N$).

Conversely, under these conditions expression (2.6), considered as a functional in R_0^N, is a characteristic functional of some (unique) Gaussian distribution on R^N.

PROOF. It is easy to verify that expression (2.6), as a functional in R_0^N, satisfies all the conditions of the theorem of Subsection 1.2.2. Therefore there exists a unique distribution μ on R^N for which the functional (2.6) is a characteristic functional. This distribution is Gaussian because an arbitrary linear continuous functional $f(x)$ has, by relation (2.4) in Subsection 2.1.3, a Gaussian distribution on the real line. The expectation a_f and the variance σ_f^2 of the random variable $f(x)$ are given by formulas

$$a_f = \sum_k f_k a_k, \qquad \sigma_f^2 = \sum_{k,j} s_{kj} f_k f_j. \tag{2.7}$$

Conversely, the characteristic functional of each normal distribution μ on R^N has the form (2.6) with the properties of the parameters given in the theorem. Indeed, from the definition of the normal distribution on R^N, for an arbitrary $f \in R_0^N$, using again relation (2.4), we have

$$\chi(tf) = \exp\left(ia_f t - \frac{1}{2}\sigma_f^2 t^2\right). \tag{2.8}$$

Now, taking into account the formula $f(x) = \sum f_k x_k$, and the relations that determine the expectation and the variance, we obtain the formulas (2.7) for a_f and σ_f^2. Then, it is sufficient to substitute these expressions into (2.8) and put $t = 1$. □

Let us now note that each component (coordinate) x_j has a Gaussian distribution and the numbers a_j and s_{kj} in the expression (2.6) are, respectively, the expectation of the Gaussian random variable x_j and the correlation of the Gaussian random variables x_k and x_j (particularly, the diagonal elements are the variances of the corresponding components). Further, as shown by the expression (2.6), the independence of the set of these components (i.e., the mutual independence of an arbitrary finite set) is equivalent to the covariance matrix being diagonal. In particular, the noncorrelation of the components of a Gaussian vector is equivalent to their independence. However, this does not mean that the noncorrelation of Gaussian random variables is equivalent to their independence (see the example at the end of this section).

2.1.6

Thus we see that a Gaussian distribution on R^n is fully characterized by two parameters: the expectation $\{a_j\}$ (the sequence of expectations of the coordinates) and the covariance matrix $\|s_{ij}\|$ (the matrix of correlations of pairs of coordinates). Knowing these parameters, one can find the probabilities for the random variable $f(x)$ to take values in a set on the real line. For this, one has to integrate over the corresponding set the one-dimensional Gaussian density with parameters a_f and σ_f, connected with the parameters of the initial Gaussian distribution by formulas (2.7). Moreover, one can also find the probabilities connected with the joint distribution of an arbitrary number of linear continuous functionals as follows: The distribution of the vector $(f^{(1)}(x), \ldots, f^{(n)}(x))$ on R^n is Gaussian with expectation $(a_f(1), \ldots, a_f(n))$ and covariance matrix $\|T_{ij}\|$, where

$$T_{ij} = \sum_{k,l} s_{kl} f_i^{(k)} f_j^{(l)} \qquad (i, j = 1, 2, \ldots, n).$$

We shall not give the proof here since it is lengthy and would essentially be a repetition of Subsection 2.1.3. By replacing the relation (2.4) by

$$\chi^{f^{(1)}, \ldots, f^{(n)}}(t_1, \ldots, t_n) = \chi(t_1 f^{(1)} + t_2 f^{(2)} + \cdots + t_n f^{(n)}),$$

we find that the meaning of the above does not need an explanation, as it follows easily from the definition of the characteristic functional.

In this way we can also easily show that this fact is true not only for the space R^N but also in the general case.

2.1.7

We shall now give an example showing that in the previous definition the requirement that all continuous linear functionals be Gaussian cannot be replaced, even in the two-dimensional case, by the condition that only the "coordinate" linear functionals (the components of the vector) be Gaussian.

Let

$$\varphi(x_1, x_2) = (2\pi)^{-1} e^{-1/2(x_1^2 + x_2^2)} \left[1 + \psi(x_1)\psi(x_2) \right],$$

where ψ is any integrable odd function, whose absolute value does not exceed one. It is obvious that the function $\varphi(x_1, x_2)$ is the density of the distribution of some non-Gaussian vector, whose components are Gaussian random variables.

Adding the condition $\int_{-\infty}^{+\infty} x\psi(x)\, dx = 0$ on $\psi(x)$, we obtain an example of a pair of Gaussian random variables that are noncorrelated but which are not independent. Also, this example shows that the sum of two Gaussian variables is not necessarily Gaussian, for if the sum of Gaussian variables were always a Gaussian variable, then an arbitrary linear combination of Gaussian variables would be a Gaussian variable, too, that is, a vector with Gaussian components would always be Gaussian.

2.2 Gaussian Distributions on the Space l_p $(1 \leqslant p < \infty)$

2.1.1

As already mentioned, measurable sets of the subspace X of the space R^N are also measurable when considered as subsets of R^N. Thus if a distribution on R^N is concentrated on X, then it is also a distribution on the space X. The converse is also true: Each distribution μ on X can be treated as the restriction of a distribution $\tilde{\mu}$ on R^N, which is concentrated on X. The considerations of Subsection 2.1.4 further show that the distribution μ on X is Gaussian if and only if the corresponding distribution $\tilde{\mu}$ on R^N is Gaussian. From this it follows that we would obtain all possible Gaussian distributions on X if we could find all possible Gaussian distributions on R^N that are concentrated on X. We shall now find Gaussian distributions on the spaces l_p.

Lemma. *Let $\xi_1(\omega), \xi_2(\omega), \ldots$ be an arbitrary sequence of random variables and $\alpha_1, \alpha_2, \ldots$ a sequence of positive numbers that increase monotonically* *to infinity. If*

$$P\{\omega : \xi_n(\omega) \geqslant \alpha_n\} \geqslant \text{const.} > 0,$$

then the sequence $\xi_1(\omega), \xi_2(\omega), \ldots$ *cannot be bounded with probability* 1, *and, in particular, it cannot, almost surely, be convergent.*

PROOF. Denote by A_n the event $\{\omega : \xi_n(\omega) \geqslant \alpha_n\}$. It is obvious that if for some $\omega = \omega_0$ the inequaltiy $\xi_n(\omega_0) \geqslant \alpha_n$ holds for infinitely many values of n, then the numerical sequence $\xi_1(\omega_0), \xi_2(\omega_0), \ldots$ is not bounded. Therefore the probability that the sequence $\xi_1(\omega), \xi_2(\omega), \ldots$ is unbounded is not less than the probability of the upper limit of the sequence of events A_1, $A_2 \ldots$. But

$$P\Big(\lim_{n\to\infty} \sup A_n\Big) = P\Big(\bigcap_{n=1} \bigcup_{k=n} A_k\Big) = \lim_{n\to\infty} P\Big(\bigcup_{k=n} A_k\Big) \geqslant P(A_n),$$

and according to the assumptions of the lemma, $P(\lim_{n\to\infty} \sup A_n) > 0$. This ends the proof of the lemma. □

2.2.2

Theorem. *If the Gaussian distribution* μ *is concentrated on* l_p $(1 \leqslant p < \infty)$, *then* $\{a_i\} \in l_p$, *where, as before,* $a_j = E_\mu x_j$, $j = 1, 2, \ldots$.

PROOF. Assume the contrary, that is, $\{a_j\} \notin l_p$. Then, as known, one can find an element f in the conjugate space $l_q(p + q = pq)$, so that the series $\sum a_k f_k$ diverges. Changing, if necessary, the signs of the corresponding coordinates of the element f, one can take all the terms of this series to be positive, and therefore the sequence of partial sums of this series increases monotonically to $+\infty$. Consider the functional $f^{(n)}(x) = \sum_{k=1}^n f_k x_k$. This functional is linear and continous and therefore has a Gaussian distribution. The expectation of this functional equals the sum $\sum_{k=1}^n f_k a_k$. The obvious (because of the normality of the distribution) equalities

$$\mu\left\{x : \sum_{k=1}^n f_k x_k \geqslant \sum_{k=1}^n f_k a_k\right\} = 1/2 \qquad (n = 1, 2, \ldots)$$

show that, according to the lemma of Subsection 2.2.1, the series $\sum f_k x_k$ should diverge on a set with positive probability. This is impossible because $x \in l_p$ almost surely and $f \in l_q$. This contradiction proves the theorem. □

The theorem of Subsection 2.2.2 gives a necessary condition for the Gaussian distribution on R^N to be concentrated on l_p. A second necessary condition is given in the next theorem, but first a simple lemma will be proved.

2.2.3

Lemma. *Let $\xi(\omega)$ be a Gaussian random variable on the probability space $(\Omega, \mathfrak{B}, P)$, with expectation m and variance σ^2. Then, for each $t \geqslant 0$ the following relation is true.*

$$\int_{\Omega} |\xi(\omega) - m|^t P(d\omega) = c(t)\sigma^t$$

where

$$c(t) = 2^{t/2}\pi^{-1/2}\Gamma((1+t)/2)$$

and Γ is Euler's gamma function.

PROOF. We have

$$\int_{\Omega} |\xi(\omega) - m|^t P(d\omega) = \sigma^{-1}(2\pi)^{-1/2}\int_{-\infty}^{+\infty} |u - m|^t e^{-(u-m)^2/2\sigma^2} du$$
$$= (2/\pi)^{1/2}\sigma^{-1}\int_0^{\infty} v^t e^{-v^2/2\sigma^2} dv;$$

and substituting in the last integral $v^2/2\sigma^2 = y$, we obtain the desired relation. □

2.2.4

Theorem. *If the Gaussian distribution μ is concentrated on l_p, $(1 \leqslant p < \infty)$, then $\{s_{kk}\} \in l_{p/2}$, where $s_{kk} = E_\mu(x_k - E_\mu x_k)^2$.*

PROOF. Denote by A_n $(n = 1, 2, \ldots)$ the following events.

$$\left\{x : \sum_{k=1}^{n} |x_k - a_k|^p \geqslant \alpha(1-\beta)s^{(n)}\right\},$$

where

$$s^{(n)} = \sum_{k=1}^{n} s_{kk}^{p/2}, \qquad \alpha > 0, \quad 0 < \beta < 1$$

and a_k is the same as before $(a_k = E_\mu x_k)$.

The proof of the theorem will be complete if we show that

$$\mu(A_n) \geqslant \text{const.} > 0 \qquad (n = 1, 2, \ldots) \tag{2.9}$$

for some values of α and β satisfing the condition $\alpha(1-\beta) > 0$.

Actually, if the sequence $\{s^{(n)}\}$ diverges, then from inequality (2.9) we obtain, according to the lemma of Subsection 2.2.1, that the series $\sum |x_k - a_k|^p$ diverges on a set of positive probability. But this is impossible because $\{a_k\} \in l_p$ almost surely according to the assumption.

Therefore we shall show that inequality (2.9) is satisfied by a suitable choice of the numbers α and β. From the definition of the events A_n, it can easily be seen that

$$\mu(A_n) \geqslant \mu\left\{x : \left| \sum_{k=1}^{n} |x_k - a_k|^p - \alpha s^{(n)} \right| \leqslant \alpha\beta s^{(n)}\right\};$$

and using Chebyshev's inequality, we obtain

$$\mu(A_n) \geqslant 1 - \frac{1}{\alpha^2\beta^2 s^{(n)2}} \left[\sum_{j,k=1}^{n} \int |x_j - a_j|^p \cdot |x_k - a_k|^p \mu(dx) \right.$$
$$\left. - 2\alpha s^{(n)} \sum_{j=1}^{n} \int |x_j - a_j|^p \mu(dx) + \alpha^2 s^{(n)2} \right].$$

Evaluating each of the terms under the double sum, using Hölder's inequality, and applying the lemma of Subsection 2.2.3, we obtain

$$\mu(A_n) \geqslant 1 - \frac{1}{\alpha^2\beta^2}\left[c(2p) - 2\alpha c(p) + \alpha^2\right].$$

Now, it is sufficient to show the existence of the values $\alpha > 0$ and $0 < \beta < 1$, for which the following inequality is satisfied.

$$\frac{1}{\alpha^2\beta^2}\left[c(2p) - 2\alpha c(p) + \alpha^2\right] < 1.$$

Let α satisfy the condition, $c(2p) - 2\alpha c(p) < 0$. Then,

$$\alpha^{-2}\left[c(2p) - 2\alpha c(p) + \alpha^2\right] < 1,$$

and β can be taken as the square root of any positive number that is strictly between $\alpha^{-2}[c(2p) - 2\alpha c(p) + \alpha^2]$ and one.

This ends the proof of the theorem. □

Corollary. *From this theorem one can immediately conclude, by applying the lemma of Subsection 2.2.3, that in the space l_p the p-th power of the norm of an element can be integrated with respect to an arbitrary Gaussian distribution. From this a stronger result can be proved, namely, the integrability of an arbitrary positive power of the norm with respect to an arbitrary Gaussian distribution (the proof of this is given in Subsection 2.3.3).*

2.2.5

Now, we shall show that the necessary conditions given in the last two theorems are also sufficient for $\mu(l_p) = 1$. To do this we shall use mainly the sufficient condition (1) of the theorem of Subsection 1.3.4. It is easy to verify that our assumptions imply this condition. One needs only to change the variable in the integral and then to use the inequality $|\alpha + \beta|^p \leqslant 2^p (|\alpha|^p + |\beta|^p)$.

Thus we have proved the following theorem.

Theorem. *In order for the Gaussian distribution on R^N with parameters $\{a_k\}$ and $\|s_{kj}\|$ to be concentrated on l_p $(1 \leqslant p < \infty)$, it is necessary and sufficient that*

$$\{a_k\} \in l_p, \qquad \{s_{kk}\} \in l_{p/2}. \tag{2.10}$$

2.2.6

Let us reformulate the previous theorem as a theorem on the general form of the characteristic functionals of all Gaussian distributions on l_p.

Theorem. *The characteristic functional of an arbitrary Gaussian distribution on the space l_p $(1 \leqslant p < \infty)$ has the form*

$$\chi(f) = \exp\left(i \sum_{k=1}^{\infty} f_k a_k - \frac{1}{2} \sum_{k,j=1}^{\infty} s_{kj} f_k f_j \right), \qquad f \in l_q,$$

where the sequence $\{a_k\}$ and the infinite symmetric nonnegative-definite matrix $\|s_{kj}\|$ satisfy the conditions (2.10). *Conversely, if these conditions are satisfied, then this formula determines the characteristic functional of some (unique) Gaussian distribution on l_p.*

2.2.7

We shall give one more formulation of the basic theorem of Subsection 2.2.5. The problem investigated here will be studied in more detail (although from a different point of view) in the last chapter (Section 4.4).

Let $\mathfrak{R}_2(l_p)$ denote the family of all real symmetric nonnegative-definite matrices $R = \|r_{ij}\|$ $(\sum r_{ij} f_i f_j \geqslant 0$ for all $f \in R_0^N)$ satisfying the condition

$$\sum_{k=1}^{\infty} r_{kk}^{p/2} < +\infty;$$

and let us induce with this family a topology $\mathfrak{T}_p$ in the space l_q (we recall that $q = p(p-1)^{-1}$ and, in particular, $q = \infty$ if $p = 1$). As the base of neighborhoods of zero in the topology $\mathfrak{T}_p$, take the family of sets[3]

$$\mathcal{O}_R(0) = \left\{ f : f \in l_q, \quad \sum_{i,j=1} r_{ij} f_i f_j < 1 \right\}, \quad R \in \mathfrak{R}_2(l_p).$$

One can show that this class of sets really gives the base of neighborhoods of zero of the locally convex Hausdorff topology $\mathfrak{T}_p$ of the linear space l_q. We shall omit the proof since it follows from more general assertions (Section 4.4).

We shall now show that the functionals

$$a(f) = \sum_{k=1}^{\infty} a_k f_k, \qquad (Sf, f) = \sum_{i,j=1}^{\infty} s_{ij} f_i f_j$$

are well defined everywhere in the space l_q and are continuous in the topology $\mathfrak{T}_p$ if and only if the conditions $\{a_k\} \in l_p$ and $\{s_{kk}\} \in l_{p/2}$ are satisfied.

If $a(f)$ is defined everywhere in l_q, then $\{a_k\} \in l_p$. This fact is well known. Conversely, let $a \in l_p$. Then, $a(f)$ is defined everywhere in l_q. Moreover, $a(f)$ is continuous in the $\mathfrak{T}_p$-topology. Indeed, let $\epsilon > 0$ be arbitrary, and denote by R_ϵ the matrix with elements

$$r_{ij}(\epsilon) = a_i a_j / \epsilon^2 \qquad (i, j = 1, 2, \ldots).$$

It is obvious that $R_\epsilon \in \mathfrak{R}_2(l_p)$; and if $(R_\epsilon f)(f) < 1$, then $|a(f)| < \epsilon$. Because ϵ is arbitrary, $a(f)$ is continuous.

Now, let us consider the functional $(Sf)(f)$. If the condition $\{s_{kk}\} \in l_{p/2}$ is satisfied, then the continuity of this functional is obvious. For an arbitrary $\epsilon > 0$ R_ϵ can be defined as the matrix $S/\epsilon \in \mathfrak{R}_2(l_p)$. Conversely, let this functional be continuous. Then, for an arbitrary $\epsilon > 0$, in particular, for some fixed $0 < \epsilon < 1$, there exists a matrix $R_\epsilon \in \mathfrak{R}_2(l_p)$ such that from $(R_\epsilon f)(f) < 1$ it follows that $(Sf)(f) < \epsilon$. If $r_{kk}(\epsilon) = 0$ for some k, then, $s_{kk} = 0$ for the same k (otherwise the element f would be taken, with the kth coordinate large and the other coordinates equal to zero and we would have $(R_\epsilon f)(f) = 0$ with $(Sf)(f) > \epsilon$). Therefore we assume that $r_{kk}(\epsilon) > 0$, $k = 1, 2, \ldots$. Let

$$f = e^{(i)} \epsilon [r_{ii}(\epsilon)]^{-1/2}.$$

---

[3]Note the fact that the matrix R is nonnegative definite in the usual sense implies the condition $|r_{ij}|^2 \leqslant r_{ii} r_{jj}$, $(i, j = 1, 2, \ldots)$; and this together with condition (2.10) shows that the expression $\sum r_{ij} f_i f_j$ is well defined and has finite value for all $f \in l_q$.

Then: $(R_\epsilon f)(f) = \epsilon^2 < 1$. Therefore $(Sf)(f) = s_{ii}\epsilon^2 / r_{ii}(\epsilon) < \epsilon$; that is, $s_{ii} < (1/\epsilon) r_{ii}(\epsilon)$ and $S \in \mathfrak{R}_2(l_p)$, because $R_\epsilon \in \mathfrak{R}_2(l_p)$.

Thus the following theorem is proved.

Theorem. *For the functional*

$$\chi(f) = \exp\left\{ if(a) - \frac{1}{2}(Sf)(f) \right\}, \qquad f \in l_q$$

to be the characteristic functional of some distribution on l_p, *it is necessary and sufficient that* $\chi(f)$ *be continuous in the* $\mathfrak{T}_p$*-topology.*

It is obvious that under the conditions of this theorem the corresponding distribution is Gaussian and determined uniquely.

2.2.8

If some distribution μ satisfies the condition

$$E_\mu x_j^2 < +\infty \qquad \text{for all } j, \quad (j = 1, 2, \ldots),$$

where E_μ means, as before, integration with respect to the measure μ over the whole space, then one can define the sequence of expectations and the covariance matrix for μ, as was done for Gaussian distributions.

Assume for simplicity that the expectations are equal to zero, and retain the notation $S = \|s_{ij}\|$ for the covariance matrix of the measure μ:

$$s_{ij} = E_\mu x_i x_j, \qquad (i, j = 1, 2, \ldots). \tag{2.11}$$

For the covariance matrix of a distribution on the space l_p it is interesting to know whether the condition

$$\sum_{k=1}^{\infty} s_{kk}^{p/2} < +\infty \tag{2.12}$$

is fulfilled, which is necessary for Gaussian distributions but is not necessary in the general case. In applications it is also interesting to know the rate of convergence of this series when the series is convergent. We shall now investigate these problems.

In the case of the space l_2, condition (2.12) is equivalent to the integrability with respect to μ of the square of the norm. The elementary Hölder's inequality and the theorem about termwise integration of a series with nonnegative functions show that in the space l_p for $p \geqslant 2$, from the integrability of the pth power of the norm, condition (2.12) follows; whereas for $p \leqslant 2$ the converse holds. From condition (2.12) the integrability of the pth power of the norm follows. The converse statements are not true if $p \neq 2$. The following example shows that for $p > 2$ condition (2.12) does not imply integrability of any positive power of the norm, and for $p < 2$, concentration of the measure on a bounded set (in particular, integrability of an arbitrary positive power) does not imply condition (2.12).

EXAMPLE. For all $k = 1, 2, \ldots$ let $\alpha_k > 0$, $\mu_k > 0$, and $\sum_{k=1}^{\infty} \mu_k = 1$. Let us take a discrete distribution μ on the space l_p, assigning the measure $(1/2)\mu_k$ to the element $\alpha_k e^{(k)}$ and $(1/2)\mu_k$ to the element $-\alpha_k e^{(k)}$ (recall that $e_j^{(k)} = 0$ for $k \neq j$ and $e_k^{(k)} = 1$). It is obvious that $E_\mu x_j = 0$ for all j, and taking into account relation (2.11), we obtain

$$\sum_{k=1}^{\infty} s_{kk}^{p/2} = \sum_{k=1}^{\infty} \alpha_k^p \mu_k^{p/2}. \tag{2.13}$$

Similarly, we can easily obtain the following.

$$E_\mu \|x\|^t = \sum_{k=1}^{\infty} \alpha_k^t \mu_k, \qquad t > 0. \tag{2.14}$$

If we first set $p > 2$ and take, for example,

$$\mu_k = k^{-1}(\ln k)^{-2}, \qquad \alpha_k = k^{(p-2)/2p},$$

then one can see that the series (2.13) converges, whereas the series (2.14) diverges for each $t > 0$. Similarly, for $p < 2$, take $\alpha_k = 1$ for all k and μ_k so that the series $\sum_{k=1}^{\infty} \mu_k^{p/2}$ diverges. Then, $\mu\{x : \|x\| \leqslant 1\} = 1$, but condition (2.12) will not be satisfied.

2.2.9

The last statement, together with the necessary part of the theorem of Subsection 2.2.6, gives a negative answer to a problem posed by R. Fortet and E. Mourier [21]. They showed[4] that if X is a Banach space of some type (determined by them) that includes, in particular, l_p for $p \geqslant 2$, then the expression

$$\chi(f) = \exp\left(-\frac{1}{2} E_\mu\left[f^2(x)\right]\right), \qquad f \in X^* \tag{2.15}$$

is the characteristic functional of some (Gaussian) distribution on X if the arbitrary distribution μ, which appears in this expression, satisfies the

[4] We use our notation and terminology.

conditions

$$E_{\mu} f(x) = 0 \quad \text{for each } f \in X^*, \quad \text{and} \quad E_{\mu}\|x\|^2 < +\infty. \qquad (2.16)$$

Is this statement true in general if no restrictions are put on the Banach space X? The answer is negative. Take as X any of the spaces l_p for $1 \leqslant p < 2$ and let μ be the distribution given in the previous subsection for $p < 2$. This distribution satisfies conditions (2.12), but not condition (2.13), and therefore expression (2.15), according to the theorem of Subsection 2.2.6, cannot be the characteristic functional of any distribution on the space l_p. Moreover, the distribution μ is concentrated on the unit ball, and thus each function of the norm is integrable with respect to μ, so that the condition $E_{\mu}\|x\|^2 < +\infty$ is not sufficient in the general case but is one of many possible conditions required for sufficiency.

2.2.10

The same examples also show that the problem of the rate of decrease of the sequence $\{s_{kk}\}$ is not connected with the integrability of the power of the norm. No rate of the convergence of the sequence s_{kk} can be concluded, even if the distribution is concentrated on a bounded set.

This statement is obviously true, not only in the case $p < 2$ but also for an arbitrary space l_p, $1 \leqslant p < \infty$.

2.3 Gaussian Distributions on the Spaces l_p ($1 \leqslant p < \infty$): Continuation

2.3.1

In this subsection we shall prove some additional facts about Gaussian distributions on the spaces l_p. These facts are related to different aspects of the theory of probability distributions and their applications. They are treated jointly because most of the properties follow as corollaries to the results proved in Section 2.2 or as examples that illustrate these results.

Let us start with problems that are traditional in the theory of probability.

By simply applying the theorem of Subsection 2.2.6, together with a theorem of A. Tortrat [30], one can obtain the general form of the characteristic functionals of all infinitely divisible distributions on the spaces l_p ($1 < p < \infty$). We shall not give the expression because it would require a number of new definitions and additional notation which will not be used elsewhere in this book. We shall go to the next problem, namely, a limit theorem for normed sums of random elements with values in R^N or l_p. One can treat this problem in terms of distributions, but the notion of a random element is more natural here.

2.3.2

Let $x_1, x_2, \ldots$ be a sequence of independent, identically distributed random elements with values in R^N. Assume that these random elements are second order in the weak sense, that is, the following condition is satisfied.

$$E\big[f(x_1)\big]^2 < +\infty \qquad \text{for all } f \in R_0^N.$$

This condition, in particular, implies the existence of the expectation of the random variables $f(x_k(\omega))$. Assume for brevity that these expectations are zero for all $f \in R_0^N$, and consider the normalized sums usually considered in the theory of probability, that is,

$$y_n = n^{-1/2}(x_1 + x_2 + \cdots + x_n). \tag{2.17}$$

Further, we let $\chi_n(f) = Ee^{if(y_n)}$ denote the characteristic functional of the random element y_n. It is obvious that $\chi_n(f) = \chi_n^f(1)$, where $\chi_n^f(t)$ is the characteristic function of the real random variable $\eta_n^f = f(y_n)$. It is also obvious that

$$\eta_n^f = n^{-1/2}(\xi_1^f + \xi_2^f + \cdots + \xi_n^f),$$

where $\xi_k^f = f(x_k)$ $(k = 1, 2, \ldots)$ is the sequence of mutually independent identically distributed real random variables with expectations zero and finite variances. Therefore, according to the classical one-dimensional limit theorem, the distributions of the random variables η_n^f are convergent to the Gaussian distribution with parameters $(0, \sigma_f^2)$, where $\sigma_f^2 = E[f(x_1)]^2$. From this we easily obtain

$$\lim_{n\to\infty} \chi_n(f) = \exp\Big(-\frac{1}{2}\sum_{i,j} s_{ij} f_i f_j\Big), \qquad f \in R_0^N, \tag{2.18}$$

where

$$s_{ij} = Ex_{1,i}x_{2,j} \qquad (i, j = 1, 2, \ldots); \tag{2.19}$$

and therefore, according to the theorem of Subsection 1.2.8, we have the result on the convergence of the sequence of normed sums, (2.17) to the Gaussian distribution on R^N, with zero mean and the covariance matrix given by (2.19).

Suppose that the limit Gaussian distribution is concentrated on l_p. For this, according to the theorem of Subsection 2.2.5, it is necessary and sufficient that the following condition be satisfied.

$$\sum_{k=1}^{\infty} \sigma^p(x_1) < +\infty, \qquad \sigma^2(x_{1,k}) = Ex_{1,k}^2. \tag{2.20}$$

If $p \leqslant 2$, then it follows from this condition that the random element x_1 (and consequently, each of the other x_k) is also concentrated on l_p (one should apply Hölder's inequality). Moreover, the theorem of Subsection 1.3.6, of which the assumptions can easily be concluded from condition (2.20) for $p \leqslant 2$, shows that the sequence of distributions of random elements (2.17) is relatively compact in the sense of the weak convergence in l_p. Thus, taking into account the convergence in R^N, we obtain the following theorem.

Theorem. *Let the elements* x_k $(k = 1, 2, \ldots)$ *assume values in the space* l_p, $1 \leqslant p \leqslant 2$, *and let condition* (2.20) *be satisfied. Then for* $n \to \infty$ *the sequence of distributions of the normalized sums* (2.17) *is weakly convergent in* l_p *to the Gaussian distribution with the characteristic functional given by* (2.18).

For $p > 2$, it does not follow from condition (2.20) that the random elements x_k are concentrated in l_p. Even if this requirement is satisfied, nothing more can be concluded than the convergence of the distributions of normalized sums (2.17) in the sense of weak convergence in l_p (see Subsection 1.4.4).

2.3.3

As a corollary to the theorem of Subsection 2.2.4 we note that an arbitrary Gaussian distribution on l_p has an absolute moment of order p, that is, $E_\mu \|x\|^p < +\infty$. Here we shall strengthen this result, showing the existence of the absolute moment of an arbitrary order. From this fact we shall derive in Subsection 2.3.4 the exponential integrability of the square of the norm.

Theorem. *Let* μ *be an arbitrary Gaussian distribution on the space* l_p, $1 \leqslant p < \infty$. *Then, for each* $t > 0$

$$E_\mu \|x\|^t < +\infty.$$

PROOF. According to the theorem of Subsection 2.2.2 and the elementary inequality $\|x\| \leqslant \|x - a\| + \|a\|$, it is sufficient to prove the integrability of $\|x - a\|^t$, where $a = \{a_j\}$, $a_j = E_\mu x_j$. Therefore for simplification of the notation we shall assume, without loss of generality, that $a = 0$. We shall also assume that the number t is a multiple of p, $t = Np$. For an arbitrary natural number n we have, using the formula for the integer power of a polynomial,

$$I_{N,n} \equiv \int_{l_p} \left(\sum_{j=1}^{n} |x_j|^p \right)^N \mu(dx)$$
$$= \sum B_{k_1,k_2,\ldots,k_n} \int_{l_p} |x_1|^{k_1 p} |x_2|^{k_2 p} |x_n|^{k_n p} \mu(dx),$$

where

$$B_{k_1,k_2,\ldots,k_n} = \frac{N!}{k_1!\,k_2!\ldots k_n!}$$

is the corresponding polynomial coefficient and the summation is over all such complexes $(k_1, k_2, \ldots, k_n)$, for which the nonnegative integers k_1, $k_2, \ldots k_n$ satisfy the condition

$$\sum_{i=1}^{n} k_i = N.$$

We shall use Hölder's inequality for the product of the functions $|x_1|^{k_1 p}$, $|x_2|^{k_2 p}, \ldots, |x_n|^{k_n p}$ and for the exponents

$$\alpha_1 = \frac{N}{k_1}, \alpha_2 = \frac{N}{k_2}, \ldots, \alpha_n = \frac{N}{k_n}, \qquad \sum_{i=1}^{n} \frac{1}{\alpha_i} = 1.$$

Applying this inequality to the right side of the previous inequality and applying the lemma of Subsection 2.2.3, we obtain

$$I_{N,n} \leqslant \sum B_{k_1,k_2,\ldots,k_n} \prod_{i=1}^{n} \left[\int_{l_p} |x_i|^{Np} \mu(dx) \right]^{k_i/N}$$
$$= c(Np) \sum B_{k_1,k_2,\ldots,k_n} \prod_{i=1}^{n} s_{ii}^{k_i p/2}.$$

Using once more (in the opposite direction) the expression for the integer power of a polynomial, we obtain the inequality

$$I_{N,n} \leqslant c(Np) \left(\sum_{k=1}^{n} s_{kk}^{p/2} \right)^N,$$

which gives, noting the possibility of integration of an increasing sequence of measurable functions,

$$E_\mu \|x\|^{Np} \leqslant c(Np) \left(\sum_{k=1}^{\infty} s_{kk}^{p/2} \right)^N, \qquad N = 1, 2, \ldots. \tag{2.21}$$

Now, we only use the theorem of Subsection 2.2.5, according to which the series on the right-hand side of inequality (2.21) is convergent for an arbitrary Gaussian distribution on l_p. □

2.3.4

Using relation (2.21) we shall now obtain an exponential estimate for the values of the Gaussian measures of complements of balls in the space l_p. According to Chebyshev's inequality, we have for an arbitrary $r > 0$ and natural N the following relation.

$$\mu\{x : \|x\| \geqslant r\} = \mu\{x : \|x\|^{Np} \geqslant r^{Np}\} \leqslant r^{-Np} E_\mu \|x\|^{Np}.$$

Putting

$$\left(\sum_{k=1}^{\infty} s_{kk}^{p/2} \right)^{2/p} = \sigma_\mu^2$$

and using the formula for $c(t)$ (see Subsection 2.2.3), we obtain, according to the previous relations, the following inequality.

$$\mu\{x : \|x\| \geqslant r\} \leqslant \left(\frac{\sigma_\mu}{r} \right)^{Np/2(Np+1)/2} \Gamma\left(\frac{Np+1}{2} \right) \pi^{-1/2}.$$

Using the known (see, e.g., reference 31, p. 253) expression for the gamma function

$$\Gamma(t) = t^{t-1/2} e^{-t} (2\pi)^{1/2} e^{\theta/12t}, \qquad t > 0, \quad 0 < \theta < 1$$

and choosing in the proper way the number N, we can obtain from inequality (2.21)

$$\mu\{x : \|x\| \geqslant r\} \leqslant K e^{-r^2/2\sigma_\mu^2},$$

where K does not depend on r (and does not depend on the distribution).

The following result is a simple consequence of this inequality.

Theorem. *Let μ be an arbitrary Gaussian distribution on the space l_p, $1 \leqslant p < \infty$. Then, we have*

$$\int_{l_p} e^{\lambda \|x\|^2} \mu(dx) < +\infty$$

for any

$$\lambda < \frac{1}{2\sigma_\mu^2}.$$

2.3.5

Here we shall prove a property of Gaussian distributions, which is equivalent to nondegeneracy.

A distribution on the finite-dimensional space R^n is called *nondegenerate* if there does not exist a hyperplane in R^n that supports this distribution or, equivalently, if the inner product (f, x) for $0 \neq f \in R^n$ is not a function that would be a constant almost everywhere in R^n. According to this definition, it is natural to call the distribution μ on the space l_p *nondegenerate* if each nonzero linear continuous functional has a nondegenerate (not concentrated at a point) distribution on the real line. Under the assumption that the covariance matrix does exist, this condition means that it is positive definite.

It is easy to show that the distribution μ is nondegenerate if the μ-measure of an arbitrary ball with a nonzero radius is positive. Indeed, if μ is degenerate, then there exists a functional $f_0 \in l_q$, such that $f_0 \neq 0$ and the set of $x \in l_p$ for which $f_0(x) = E_\mu f_0(x)$ has the full measure. But this set is closed (as the preimage of a one-point set) and is not identical with l_p (because $f_0 \neq 0$). Then, the complement of this set would be a nonvoid open set with zero measure, whose existence contradicts the assumption.

Now, we shall show that for Gaussian distributions on l_p, $1 \leqslant p < \infty$, the converse is also true.

First, note that the covariance matrix S of a Gaussian distribution on l_p induces a linear bounded mapping $l_q \to l_p : (Sf)_i = \sum_{j=1}^{\infty} s_{ij} f_j$ $(i = 1, 2, \ldots)$. The boundedness of S is easy to verify using the property $s_{ij}^2 \leqslant s_{ii} s_{jj}$, Hölder's inequality, and the theorem of Subsection 2.2.5 (the necessary part).

Theorem. *Let μ be an arbitrary nondegenerate Gaussian distribution on l_p $(1 \leqslant p < \infty)$. Then, the μ-measure of an arbitrary ball in l_p with a nonzero radius is positive.*

PROOF. Denote by $\mathcal{O}_r(x)$ the ball with the center at a point x and with radius r. We shall show that if there exists at least one ball $\mathcal{O}_{r_0}(x_0)$ with a nonzero radius, which has measure zero, then the measure of an arbitrary ball with the same radius and center at point $x_0 + Sa$, where a is an arbitrary finite length element $(a \in R_0^N)$, is also zero. This will complete the proof. Indeed, because the distribution is nondegenerate, $S(l_q)$ is everywhere dense in l_p (otherwise there would exist an element $f_0 \in l_q$ such that

$f_0(Sf) = 0$ for all $f \in l_q$, and, in particular, $f_0(Sf_0) = 0$, which contradicts the nondegeneracy). Further, because of the boundedness of the mapping S, the image of an arbitrary set that is everywhere dense in l_q is everywhere dense in l_p. In particular, the denumerable set of elements of the form $x_0 + Sa$ with finite lengths Sa, which have rational coordinates, is everywhere dense in l_p.[5] Therefore the condition $\mu(\mathcal{O}_{r_0}(x_0 + Sa)) = 0$ for all such a creates a contradiction (because of the countable additivity of μ) with the condition $\mu(l_p) = 1$.

So we shall show that from $\mu(\mathcal{O}_{r_0}(x_0)) = 0$ it follows that $\mu(\mathcal{O}_{r_0}(x_0 + b)) = 0$, where $b = Sa$ and a is an arbitrary finite length element. First, note that according to the condition $\{E_\mu x_j\} \in l_p$ (theorem of Subsection 2.2.2) and because x_0 is arbitrary, we can say without loss of generality that $E_\mu x_j = 0, j = 1, 2, \ldots$.

It is obvious that the ball $\mathcal{O}_{r_0}(x_0 + b)$ is the intersection over all n of the decreasing sequence of cylinder sets in l_p, which are determined by conditions

$$\sum_{k=1}^{n} |x_{0k} + b_k - x_k|^p \leqslant r_0^p. \tag{2.22}$$

Therefore because of the continuity of the measure, we have

$$\mu(\mathcal{O}_{r_0}(x_0 + b)) = \lim_{n\to\infty} \mu^{(n)}\left(\mathcal{O}_{r_0}^{(n)}(x_0^{(n)} + b^{(n)})\right), \tag{2.23}$$

where $x_0^{(n)}$ and $b^{(n)}$ are the corresponding projections of the points x_0 and b on the subspace R^n of the first n coordinates $(x_{0k}^{(n)} = x_{0k}, b_k^{(n)} = b_k, k = 1, 2, \ldots, n)$, $\mathcal{O}_{r_0}^{(n)}(x_0^{(n)} + b^{(n)})$ is a set in R^n, defined by relation (2.22), and $\mu^{(n)}$ is the projection of μ on R^n, that is, the Gaussian distribution on R^n with expectation zero and the covariance matrix $S^{(n)}$ $(S^{(n)} = \|s_{ij}^{(n)}\|, s_{ij}^{(n)} = s_{ij}, i, j = 1, 2, \ldots, n)$.

The expression $\mu^{(n)}(\mathcal{O}_{r_0}^{(n)}(x_0^{(n)} + b^{(n)}))$ can be given by the integral of the density of the distribution μ^n over the particular set. Changing variables in this integral, we obtain

$$\mu^{(n)}\left(\mathcal{O}_{r_0}^{(n)}(x_0^{(n)} + b^{(n)})\right) = \int_{\mathcal{O}_{r_0}^{(n)}(x_0^{(n)})} \Phi_n(x^{(n)})\mu^{(n)}(dx^{(n)}), \tag{2.24}$$

[5] If $q = \infty$ $(p = 1)$, then this conclusion follows from subsequent reasoning: The countable set of finite length elements with rational coordinates is everywhere dense in l_∞ with respect to the topology $\sigma(l_\infty, l_1)$. The operator S, being a symmetric map $l_\infty \to l_1$, is continuous with respect to $\sigma(l_\infty, l_1)$ in l_∞ and $\sigma(l_1, l_\infty)$ in l_1. This is easy to verify. Therefore the set of points $\{x + Sa\}$, where $a \in R_0^N$ and a has rational coordinates, is everywhere dense in l_1 with respect to the weak topology $\sigma(l_1, l_\infty)$. Since this set is convex, it is also dense with respect to the strong topology in l_1 (weak and strong closures of convex sets are the same, see, for example, reference 11, p. 422).

where

$$\Phi_n(x^{(n)}) = \exp\left\{ -(T^{(n)}b^{(n)}, x^{(n)}) - \frac{1}{2}(T^{(n)}b^{(n)}, b^{(n)}) \right\}. \qquad (2.25)$$

$T^{(n)}$ is the matrix inverse to S^n (the existence of T^n follows from the nondegeneracy of the distribution).

Now, note that, according to our assumption $b = Sa$, there exists a number $L = L(a)$, such that $a_k = 0$ for $k > L$. Therefore for $n > L$ the element $b^{(n)}$ can be represented in the form $b^{(n)} = S^{(n)}a^{(n)}$. Substituting this value of $b^{(n)}$ into (2.25), we obtain

$$\Phi_n(x^{(n)}) = \Phi_L(x^{(L)}) = \exp\left\{ -(a^{(L)}, x^{(L)}) - \frac{1}{2}(S^{(L)}a^{(L)}, a^{(L)}) \right\}.$$

Thus for $n > L$ the function Φ_n does not depend on n; and it can be considered as a functional in l_p, which does not depend on the coordinates whose indexes are greater than L. Therefore for $n > L$ the integral on the right-hand side of (2.24) can be replaced by the integral of the functional $\Phi(x) = \Phi_L(x^L)$ with respect to the measure μ over the corresponding cylinder set in l_p. Taking into account (2.23) and (2.24) and the absolute continuity of the integral, we obtain

$$\mu(\mathbb{O}_{r_0}(x_0 + b)) = \int_{\mathbb{O}_{r_0}(x_0)} \Phi(x)\mu(dx).$$

From this, because of the obvious integrability of the functional $\Phi(x)$ over the ball $\mathbb{O}_{r_0}(x_0)$ and because of the assumption $\mu(\mathbb{O}_{r_0}(x_0)) = 0$, we have $\mu(\mathbb{O}_{r_0}(x_0 + b)) = 0$. This ends the proof of the theorem. □

2.3.6

The degree of the degeneracy of the distribution can in some sense be characterized by the family of linear functionals that have degenerate distributions. One of the corollaries of the theorem of Subsection 2.2.5 is the existence, for an arbitrary $p > 1$, of nontrivial Gaussian distributions on l_p that have infinite-dimensional degeneracy and are not concentrated on any $l_{p'}$ for $p' < p$. For example, it is sufficient to take an arbitrary covariance matrix S that satisfies the required condition $\sum_{k=1}^{\infty} s_{kk}^{p/2} < +\infty$ and has infinitely many zeros; and for the sequence of expectations take a sequence that belongs to l_p but which does not belong to any $l_{p'}$ if $p' < p$.

2.3.7

In this subsection we shall give one more example of the application of the theorem of Subsection 2.2.5, but now we turn to a different kind of problem. We shall investigate the Cauchy problem for the stochastic heat conduction equation

$$\frac{\partial W(x,t)}{\partial t} - \frac{\partial^2 W(x,t)}{\partial x^2} = \delta(x) f(t), \tag{2.26}$$

which describes the process of heat conduction in a rod of length 2π, with a random point source $f(t)$, which is the derivative of the Wiener process ("white noise"), having the initial conditions

$$W(x,0) = W^0(x), \qquad -\pi \leqslant x \leqslant +\pi \tag{2.27}$$

and periodic boundary conditions, according to which, as a physical interpretation for the conductor of heat, an arbitrary closed linearly stretched conductor with the same length may be used instead of the rod.

Taking the boundary condition into account and using the Fourier transform method, we obtain the following system of stochastic equations for the random amplitudes $W_k(t)$, which are the Fourier cosine coefficients[6] of the solutions of equation (2.26):

$$\frac{dW_k(t)}{dt} + k^2 W_k(t) = f(t) \qquad (k = 0, 1, 2, \ldots), \tag{2.28}$$

with the initial condition

$$W_k(0) = W_k^0 \qquad (k = 0, 1, 2, \ldots), \tag{2.29}$$

where W_k^0 are the Fourier cosine coefficients of the function $W^0(x)$, which occur in corresponding condition (2.27).

The problem considered here has been investigated previously in reference 32, in which the author used the Fourier transform and also obtained the system of equations (2.28), and then using complicated tools he obtained the following two results.

a. The finite-dimensional distributions that correspond to solutions of the system, (2.28) can be extended for each $t > 0$ to a (Gaussian) distribution on the Hilbert space l_2,
b. Almost all solutions $\{W_k(t)\}$ are, for each $t > 0$, "fast" decreasing sequences from l_2 (they decrease faster with respect to k than any power of $1/k$).

From (b) we conclude that almost all solutions of equation (2.26) are infinitely differentiable with respect to x for all fixed $t > 0$.

---

[6] It is not necessary to investigate the sine coefficients, because they are not influenced by randomness, and therefore they are determined by the known classical formulas.

Using the theorem of Subsection 2.2.5, we shall give a simple proof of (a), which essentially does not need any calculations; and in addition, we shall show that (b) is false.

Let $\{W_k(t)\}$ be the solution of the system of equations (2.28) with the corresponding initial condition (2.29), and let $\mu_t^{(n)} = \mu_t^{(0,1,\ldots,n)}$ denote the distribution of the vector $(W_0(t), W_1(t), \ldots, W_n(t))$. It easy to see (see remark of Subsection 1.2.2), that the system of finite-dimensional distributions $\mu_t^{(n)}$ can be (uniquely) extended to a probability distribution on R^N. Denote this distribution by μ_t and note that it is Gaussian. This can be seen to be true as follows. The equations of the system (2.28) for different k are not related to each other, and the solution for a particular k is

$$W_k(t) = W_k^0 e^{-k^2 t} + \int_0^t e^{k^2(\tau - t)}\, dw(\tau),$$

where $w(\tau)$ is the Wiener process ($dw(\tau) = f(\tau)\,d\tau$), and the integral is taken in the sense of mean square convergence of integral sums. Each integral sum is a Gaussian random variable, and therefore $W_k(t)$ is, for each t, a Gaussian random variable. Similar considerations show that for each n the distribution $\mu_t^{(n)}$ on R^{n+1} is Gaussian (an arbitrary linear functional of the vector $(W_0, W_1, \ldots, W_n)$ is also the limit of Gaussian random variables), which means that μ_t is Gaussian.

The expectation of μ_t is the sequence $\{W_k^0 e^{-k^2 t}\}$, because the expectation of the increment of the Wiener process is zero. Let us find the covariance matrix s_{ij}. We have

$$s_{ij} = E\left[\int_0^t e^{i^2(\tau - t)}\, dw(\tau) \int_0^t e^{j^2(\tau - t)}\, dw(\tau)\right].$$

Substituting here the corresponding integral sums instead of the integrals and taking into account the independence of the increments of the Wiener process, we easily obtain that

$$s_{ij} = s_{ij}(t) = \frac{1 - e^{-(i^2 + j^2)t}}{i^2 + j^2} \qquad (i, j = 1, 2, \ldots);$$

and an application of the theorem of Subsection 2.2.5 immediately gives assertion (a). Moreover, the distribution μ_t is concentrated for each $t > 0$ on an arbitrary l_p, provided $p > 1$. But $\sum s_{kk}^{1/2} = +\infty$, hence μ_t is not concentrated on l_1 for any t. From this it follows that (b) is false.

Finally, we note that slightly more general linear systems can be similarly investigated. For example, on the right-hand side of the system of equations (2.28), one can take $a_k f(t)$ instead of $f(t)$, where $\{a_k\}$ is an arbitrary (real) numerical sequence. This corresponds to the case where in equation (2.26) rather than a point source there is a distributed heat source.

2.4 Gaussian Distribution on the Spaces c_0 and l_∞

2.4.1

As already shown (see Subsection 2.2.1), the characterization of all Gaussian distributions on the spaces c_0 and l_∞ is equivalent to finding all such Gaussian distributions on the space R^N that are concentrated on c_0 or on l_∞, respectively. In order to solve this problem we shall use the following lemma.

Lemma. *If* $0 \leqslant \alpha_k \leqslant 1$ *for all* $k = 1, 2, \ldots$ *then the series*[7]

$$\varphi = \sum_{k=1}^{\infty} \alpha_k \prod_{j=1}^{k-1} (1 - \alpha_j)$$

is convergent, and $\varphi \leqslant 1$. *Furthermore, for the series* $\sum_{k=1}^{\infty} \alpha_k$ *to be convergent it is necessary and sufficient that* $\varphi < 1$.

PROOF. The proof can easily be obtained by using the elementary identity

$$\prod_{j=1}^{k} (1 - \alpha_j) - \prod_{j=1}^{k-1} (1 - \alpha_j) = -\alpha_k \prod_{j=1}^{k-1} (1 - \alpha_j).$$

Summing up these identities with respect to k from 1 to N, simplifying the left-hand side of the relation, and taking the limit as $N \to \infty$, we obtain

$$\varphi = 1 - \prod_{j=1}^{\infty} (1 - \alpha_j).$$

Finally, it is enough to note that always

$$0 \leqslant \prod_{j=1}^{\infty} (1 - \alpha_j) \leqslant 1$$

with equality to zero holding if and only if $\sum_{k=1}^{\infty} \alpha_k = +\infty$. □

2.4.2

We shall now characterize Gaussian distributions. First, we consider the space c_0. Assume that μ is the Gaussian distribution on R^N with parameters $\{a_j\}$ and $\|s_{ij}\|$. Let the following conditions be satisfied.

$$\{a_j\} \in c_0, \tag{2.30}$$

$$\sum_{k=1}^{\infty} e^{-\epsilon/s_{kk}} < +\infty \qquad \text{for any } \epsilon > 0. \tag{2.31}$$

---

[7]The product $\prod_{j=1}^{0}(1 - \alpha_j)$ is defined to be equal to one.

We shall prove that under these conditions, $\mu(c_0)=1$. Note first that if $\mu\{x:\lim_{j\to\infty}(x_j-a_j)=0\}=1$, then, according to condition (2.30), we also have $\mu\{x:\lim_{j\to\infty}x_j=0\}=1$. Therefore, without loss of generality, we assume that $a_j=0$ for all j.

Writing in symbols the convergence of a sequence to zero, we obtain the following expression for the space c_0.

$$c_0=\bigcap_{\epsilon>0}\bigcup_{n=1}^{\infty}\bigcap_{k=n}^{\infty}\{x:|x_k|<\epsilon\}.$$

The intersection over all positive ϵ can be replaced by the denumerable intersection over all $\epsilon_m=1/m$ $(m=1,2,\dots)$. We shall use this fact. Taking the complement in R^N of the space c_0, we note that the measure of a denumerable union of sets is zero if and only if the measure of each set is zero. Then, we use the continuity of the measure (the measure of the intersection of a decreasing sequence of sets is equal to the limit of the measures), and it is easy to show that the required condition $\mu(c_0)=1$ is equivalent to the condition

$$\lim_{n\to\infty}\mu\left(\bigcup_{k=n}^{\infty}\{x:|x_k|\geqslant\epsilon\}\right)=0 \qquad \text{for each } \epsilon>0. \tag{2.32}$$

Let us estimate from above the measure of the union of the sets $\{x:|x_k|\geqslant\epsilon\}$ by sums of measures. We note that x_k is a Gaussian random variable with zero mean and variance s_{kk} and use the known (see, e.g., reference 33, p. 166) estimate from the preceding indefinite integral of the Gaussian density. Then, it follows that if condition (2.31) is satisfied, condition (2.32) is also satisfied, and therefore the sufficiency of conditions (2.30) and (2.31) for the distribution μ to be concentrated on c_0 is established.

Are these conditions also necessary? It is easy to show that condition (2.30) is necessary. We shall not give a detailed proof, because it is similar to the proofs of the theorem of Subsection 2.2.2 and lemma of Subsection 2.2.1 and also follows from a more general result (theorem of Subsection 4.2.3), which will be given later.

Condition (2.31) is not necessary. One can easily check this by considering a Gaussian random element $\{x_k\}\in c_0$ where $x_k=\lambda_k x$ and x is a normalized Gaussian random variable with $\lambda_k\to 0$ so slowly that condition (2.31) is not satisfied.

However, in the case where the covariance matrix is diagonal, the condition (2.31) is necessary; which we shall prove. We use the elementary expression

$$\mu\left(\bigcup_{j=n+1}^{\infty}A_j\right)=\mu(A_{n+1})+\mu(A_{n+2}A_{n+1}^c)+\mu(A_{n+3}A_{n+1}^cA_{n+2}^c)+\cdots,$$

which, under the assumption that all events under consideration are independent, can be written as

$$\mu\left(\bigcup_{j=n+1}^{\infty} A_j\right) = \sum_{k=1}^{\infty} \mu(A_{n+k}) \prod_{j=1}^{k-1} (1 - \mu(A_{n+j})). \tag{2.33}$$

Suppose that a Gaussian random element in c_0 has a diagonal covariance matrix. Without loss of generality, we assume that the expectations of the coordinates are zero and the coordinates are nondegenerate (all numbers on the diagonal are positive). The diagonality of the covariance matrix means mutual independence of all the coordinates. Therefore denoting for an arbitrary $\epsilon > 0$ the event $\{x : |x_j| \geqslant \epsilon\}$ by $A_j = A_j(\epsilon)$, we can use formula (2.33). Applying the lemma (for $\alpha_k = \mu(A_{n+k})$) and using condition (2.32), we obtain the convergence of the series $\sum_{k=1}^{\infty} \mu(A_k)$ for any $\epsilon > 0$. Finally, we use the known ([33], p. 166) estimate, previously given, for the indefinite integral of the Gaussian density.

Thus we have proved the following theorem.

Theorem. *Let* μ *be a Gaussian distribution on* R^N. *If conditions* (2.30) *and* (2.31) *are satisfied, then* $\mu(c_0) = 1$. *Conversely, if* $\mu(c_0) = 1$, *then condition* (2.30) *is satisfied, and if the covariance matrix is diagonal, then condition* (2.31) *is also satisfied.*

2.4.3

Let us now consider the space l_∞. Let μ denote, as before, the Gaussian distribution on R^N with parameters $\{a_j\}$ and $\|s_{ij}\|$ and let the following conditions be satisfied.

$$\{a_j\} \in l_\infty, \tag{2.34}$$

$$\sum_{k=1}^{\infty} e^{-r/s_{kk}} < +\infty \qquad \text{for some } r > 0. \tag{2.35}$$

We shall show that if these conditions are satisfied, then $\mu(l_\infty) = 1$. Using condition (2.34), we again assume from the beginning, without loss of generality, that $a_j = 0$ for all j. Let us write the space l_∞ as

$$l_\infty = \bigcup_{r=1}^{\infty} \bigcap_{k=1}^{\infty} \{x : |x_k| < r\}.$$

Taking the complement and using once more the continuity of the measure, we obtain the fact that the condition $\mu(l_\infty) = 1$ is equivalent to the condition

$$\lim_{r\to\infty} \mu\left(\bigcup_{k=1}^{\infty} \{x : |x_k| \geqslant r\}\right) = 0,$$

which should follow from condition (2.35). Estimating the measure of the union by the sum of the measures and using the earlier mentioned estimate of the indefinite integral of the Gaussian density, we can easily show that this is true.

The necessity of condition (2.34) follows simply from the lemma of Subsection 2.2.1, for if $\{a_j\} \notin l_\infty$, then one can assume, without loss of generality, that $a_j \to +\infty$; however, $\mu\{x : x_j \geqslant a_j\} = 1/2$ for all j, because the distribution is Gaussian.

The necessity of condition (2.35) does not hold in general. As a simple example, we can take the distribution of the sequence $(x, x, \dots)$, where x is some Gaussian random variable (with zero mean). However, by assuming that the covariance matrix is diagonal, we find that condition (2.35) is necessary. This can also be proved similarly to the corresponding statement in Subsection 2.4.2; that is, one should use formula (2.33) (assuming $n = 0$ and $A_j = A_j(r) = \{x : |x_j| \geqslant r\}$) and the lemma proved in Subsection 2.4.1 (taking $\alpha_k = \mu\{x : |x_k| \geqslant r\}$).

Thus we have proved the following theorem.

Theorem. *Let μ be a Gaussian distribution on R^N. If conditions* (2.34) *and* (2.35) *are satisfied, then $\mu(l_\infty) = 1$. Conversely, if $\mu(l_\infty) = 1$, then condition* (2.34) *is satisfied, and under the assumption of diagonality of the covariance matrix condition* (2.35) *is also satisfied.*

2.4.4

The nondegeneracy of the distribution on l_∞ can be defined in the usual way (see Subsection 2.3.5) using, however, only linear functionals $f \in l_1 \subset l_\infty^*$. For the Gaussian distribution with a diagonal covariance matrix, nondegeneracy simply means that all the numbers on the diagonal are positive.

We have shown that on the space l_p, for $1 \leqslant p < \infty$, as well as on the finite-dimensional space, each nondegenerate Gaussian distribution has the property that the measure of an arbitrary ball with a nonzero radius is positive. This property seems so natural that it might be thought that it characterizes Gaussian distributions in general and not the spaces on which they are determined. But this is not true; and l_∞ is an example of a space on which Gaussian distributions do not have this property. Moreover, we shall show that for each $r > 0$, there exists on the space l_∞ a nondegenerate Gaussian distribution μ such that the measure of an arbitrary ball with radius r is zero; that is, $\mu(\mathbb{O}_r(a)) = 0$, $a \in l_\infty$.

Let $\{\lambda_k\}$ be a sequence of positive numbers satisfying the condition

$$\sum_{k=1}^{\infty} \sqrt{\lambda_k}\, e^{-\rho/\lambda_k} < +\infty \qquad \text{for some } \rho > r^2/2 \tag{2.36}$$

and let μ be a Gaussian distribution on R^N with expectation zero and a diagonal covariance matrix with the numbers λ_k on the diagonal (the existence of μ follows from the theorem of Subsection 2.1.5). Then, μ is a nondegenerate Gaussian distribution on l_∞, because $\lambda_j > 0$ for all j, and according to the theorem of Subsection 2.4.3, $\mu(l_\infty) = 1$.[8]

Further, assume that in addition to condition (2.36) the following condition is satisfied.

$$\sum_{k=1}^{\infty} \sqrt{\lambda_k}\, e^{-r^2/2\lambda_k} = +\infty. \tag{2.37}$$

Then, $\mu(\mathcal{O}_r(a)) = 0$ for each $a \in l_\infty$. In fact, it can easily be seen that $\mu(\mathcal{O}_r(a) \leqslant \mu(\mathcal{O}_r(0))$. Further, from the elementary relation

$$\mu\left\{x : \max_{1 \leqslant j \leqslant n} |x_j| < r\right\} = \prod_{j=1}^{n} \left(1 - \mu\left\{x : |x_j| \geqslant r\right\}\right)$$

it follows that $\mu(\mathcal{O}_r(0)) = 0$, provided the infinite product on the right-hand side diverges to zero; that is, if

$$\sum_{j=1}^{\infty} \mu\left\{x : |x_j| \geqslant r\right\} = +\infty. \tag{2.38}$$

Finally, we note that condition (2.37) implies the divergence of the series (2.38). This is easy to prove using the estimate of the indefinite integral of the Gaussian density mentioned in Subsection 2.4.4.

---

[8] It is easy to see that condition (2.36) implies condition (2.35).

3

Distributions on Hilbert Space

Introduction

Results given at the beginning of Section 3.1 are particular cases of the results proved in the first two chapters for $p = 2$. Many are identical in this case with earlier known results. The theorem of Subsection 3.1.4 is the most important (a particular case of the theorem of Subsection 2.2.6), which is in our opinion the basic result of the work by E. Mourier [3],[1] which has already been mentioned.

The theorem of Subsection 3.1.5 is also new for the case $p = 2$. Further, in Subsection 3.1.6, by using the theorem of Subsection 1.2.10 for $p = 2$, we prove Sazonov's theorem, mentioned in the general introduction. There now exist several proofs of this known result (in the form of Sazonov's theorem or Minlos' theorem). For example, the proof by V. N. Sudakov is based on the theorem on the completeness of a system of generalized eigenfunctions of a self-adjoint operator.[2] The proof given here is basically the same as that given by V. V. Sazonov [16]. Some simplification is accomplished by using the method of embedding into R^N, which allows the use of countably additive measures on R^N instead of using weak distributions. Moreover, in Section 3.1 there are some results not related to those of the previous chapter for the more general case of spaces l_p and which are proved only for Hilbert spaces. In connection with this, we mention here an

---

[1] This theorem as formulated in Subsection 3.1.4, although it follows easily from the results of E. Mourier, is not given explicitly in her work. This form of the theorem has been given by Yu. V. Prohorov [4].

[2] V. N. Sudakov spoke about this problem at the International Congress of Mathematicians in Moscow in the summer of 1966.

inequality for the density of the distribution of the square of a Gaussian random element.

Section 3.2 gives estimates of the convergence rate in the central limit theorem. First, we consider the case of a finite-dimensional space. The estimate in the finite-dimensional case was recently investigated by V. V. Sazonov [34], who found some general results along this line. But in our case, by the assumption of independent components, our estimate is stronger.

Then, we turn to the infinite-dimensional case. The problem of estimating the convergence rate in a Hilbert space has been posed by Yu. V. Prohorov and investigated first by N. P. Kandelaki [35], who used the method of approximation by finite-dimensional distributions, using estimates for the finite-dimensional case and taking the limit with respect to the dimension. For one important special case Kandelaki's estimate has been strengthened by V. V. Sazonov, who also used the same method, together with an improvement of the estimate in the finite-dimensional case [36].

We use the coordinate free approach and operate directly on distributions (and their characteristic functionals) on a Hilbert space. In this way we obtain improvements and generalizations of Kandelaki's results. This method is particularly simple and natural for estimation of the convergence rate on a class of ellipsoids, which are determined by nuclear operators.

The results of this and the next section were obtained in collaboration with N. P. Kandelaki.

In Section 3.3 we investigate three problems. The first is the estimation of large deviations from the expectation of a normalized sum of independent random elements in a Hilbert space. This investigation has been motivated by an issue of the journal, *The Theory of Probability and Its Applications*, (Vol. 13, No. 2, 1968). In this issue an article by Yu. V. Prohorov was published on the construction of a finite-dimensional analogue of the well-known exponential inequality of S. N. Bernstein for the estimation of large deviations in the one-dimensional case and on the works of A. V. Prohorov and V. M. Zolotarov that preceded this investigation. Our inequality also has an exponential form. It estimates the measure of the complement of ellipsoids that are determined in the most general case by nuclear operators, and, with some assumptions, by arbitrary bounded operators. Note that our method does not require identical distributions for the summands.

Further, we investigate the distribution of the inner product of two Gaussian random elements in a Hilbert space H. The inner product can be written in the form of a quadratic functional in the Cartesian product $H \times H$, and in this way we are able to change the problem to one of finding the distribution of a quadratic functional of Gaussian random elements in a Hilbert space.

Finally, we define the integral of operator-valued functions with respect to measures with values on the Hilbert space of all random elements in H, which are strongly second order. We describe the class of integrable functions and find the expansion of the covariance operator of the integral in the form of some "nonrandom" integral in the Bochner sense. An important particular case of the measures under investigation is given by Wiener processes with values in H. Another approach to the definition of the integral in this case has been given by Yu. L. Daletskiĭ, who generalized the constructions used previously by V. V. Baklan and T. L. Chantladze (see Yu. L. Daletskiĭ [37]).

The main results of this chapter are contained in references 8, 23, 24, and 38–41.

3.1 Some General Problems

3.1.1

All the results proved in the first two chapters for the space l_p for $p=2$ hold also for an arbitrary real separable Hilbert space H, if $e_1, e_2, \ldots$ is an arbitrary basis (complete orthonormal system) in H and $x_1, x_2, \ldots$ are the coordinates of the point x in this basis. The fact that this is true follows simply from the Riesz–Fischer theorem given in Subsection 1.3.2, because all the notions used in the previously formulated results (for $p=2$) are invariant with respect to isometric isomorphisms of Hilbert spaces (Borel sets are transformed into Borel sets; measurable, continuous, linear functionals are also correspondingly transformed; Gaussian distributions are transformed into Gaussian; the norm is transformed into a norm, etc.). Accordingly, it is sufficient simply to treat $x_1, x_2, \ldots$ as the sequence of coordinates of a point with respect to some (arbitrarily chosen) basis, and all the theorems can be interpreted not only for the case of the space l_2 but also for the general case of a space H.

Therefore we shall not rewrite all the results in the case of an arbitrary space H but shall only show the most important ones, noting their relation to known results. Let us recall that a distribution (probability distribution) on H is a nonnegative normed countably additive function, determined on the σ-algebra of Borel sets in the space H. Recall also the possibility of two equivalent investigations: in terms of distributions or in terms of random elements. When speaking about a random element we imply (without explicitly stating) that there is some probability space $(\Omega, \mathfrak{B}(\Omega), P)$ and a random element with values in H that is a mapping of Ω into H, which is measurable with respect to the σ-algebra $\mathfrak{B}(H)$ of Borel sets in H and the σ-algebra $\mathfrak{B}(\Omega)$ in Ω. The random element x is characterized by its distribution μ, which is defined by

$$\mu(E) = P\{\omega : x(\omega) \in E\}, \qquad E \in \mathfrak{B}(H).$$

Conversely, each distribution μ on H is the distribution of some random element, which can be taken, for example, as the identity transformation of H into itself, choosing $(H, \mathfrak{B}(H), \mu)$ as $(\Omega, \mathfrak{B}(\Omega), P)$.

3.1.2.

The theorem of Subsection 1.3.6 for $p = 2$ is easily seen to be identical with the following theorem, due to Yu. V. Prohorov [4].

Theorem. *For the weak relative compactness of the family* $\{\mu_\alpha\}$ *of distributions on* H *it is sufficient that the following two conditions be satisfied*:

a. $\sup_\alpha \sum_{k=1}^{\infty} \int_H x_k^2 \mu_\alpha(dx) \leqslant C$,
b. $\lim_{n\to\infty} \sup_\alpha \sum_{k=n}^{\infty} \int_H x_k^2 \mu_\alpha(dx) = 0$,

where C *is a positive constant and* $x_1, x_2, \ldots$ *are the coordinates of the point* $x \in H$ *in some (arbitrary) basis.*

3.1.3

For $p = 2$ the theorem of Subsection 1.4.2 takes the following form due to V. V. Sazonov [5].

Theorem. *For the functional* $\chi(f)$, $f \in H$, *to be the characteristic functional of some distribution* μ *on* H, *such that*

$$\int_H \|x\|^2 \mu(dx) < +\infty,$$

it is necessary and sufficient that the following conditions are satisfied:

a. χ *is positive-definite.*
b. $\chi(0) = 1$.
c. χ *is continuous in the norm.*
d. $\sum_{k=1}^{\infty} \int_{-\infty}^{+\infty} u^2 \mu^{(k)}(du) < +\infty$, *where* $\mu^{(k)}$ *is the distribution on the real line whose characteristic function is* $\varphi_k(t) = \chi(te_k)$ *and* $\{e_k\}$ *is an arbitrary basis in* H.

3.1.4

Theorems of Subsections 2.2.2 and 2.2.4 are for $p = 2$ equivalent to the corresponding results of E. Mourier [3]. In the same paper we can also find one of the basic results of E. Mourier; namely, the following theorem on the general form of the characteristic functionals of all Gaussian measures on H. Recall that a distribution μ is called *Gaussian* if all linear continuous functionals (h, x) considered as random elements in $(H, \mathfrak{B}(H), \mu)$ have

Gaussian distributions. This is equivalent to the fact that an arbitrary finite set of coordinates (in an arbitrary basis) has a Gaussian distribution.

Theorem. *The characteristic functional of an arbitrary Gaussian distribution on* H *has the form*

$$\chi(h) = \exp\{i(h,a) - \tfrac{1}{2}(Sh,h)\}, \qquad h \in H, \tag{3.1}$$

where $a \in H$ *and* $S: H \to H$ *is a linear self-adjoint nonnegative completely continuous operator with a finite trace.*[3] *Conversely, if* $a \in H$ *and* S *is an operator with these properties, then expression* (3.1) *is the characteristic functional of some (unique) Gaussian distribution on* H.

This theorem is contained in the theorem of Subsection 2.2.6, which in the case of $p = 2$ is the coordinate form of Mourier's theorem, given in an invariant form (i.e., not related to a basis). This possibility of having two forms (coordinate and noncoordinate) is connected not only with Mourier's theorem but also with all problems for distributions on Hilbert spaces. The relation between the coordinate and the noncoordinate forms of representation is determined simply by fixing the basis, of which the choice is usually quite arbitrary. Sometimes, however, it is convenient to take as a basis the complete system of normalized eigenelements of some completely continuous self-adjoint operators that arise in particular problems. Most often this is the covariance operator of the Gaussian distribution with which the problem is connected. The convenience of such a choice is explained by the obvious fact that in the basis of eigenelements of the covariance operator, the matrix that corresponds to this operator is diagonal, and therefore the coordinates in this basis are mutually independent.

The element a and the operator S that occur in formula (3.1) for the characteristic functional are the expectation and the covariance operator of the distribution μ, respectively. In the coordinate form the definitions of the expectation and the covariance operator have already been given. Let us now recall these definitions in the noncoordinate form. The expectation and the covariance operator of the distribution μ (or of the random element with the distribution μ) are defined, respectively, as an element $a \in H$ and a linear bounded self-adjoint nonnegative operator S, which satisfy for all $h \in H$ the following relations.[4]

$$(h,a) = E_\mu(h,x), \qquad (Sh,h) = E_\mu(h,x-a)^2. \tag{3.2}$$

[3] Following Yu. V. Prohorov, such operators will be called *S-operators*.

[4] In the interpretation of the random elements in relations (3.2), instead of integration with respect to μ there would be integration with respect to P. A similar remark is also valid for the theorem of Subsection 3.1.5.

The existence of the expectation and the covariance operator of an arbitrary Gaussian distribution in H follows from the theorems of Subsections 2.2.2 and 2.2.4.

Recall also that the *trace* of the operator S, TrS, is defined as the sum of all diagonal elements of the corresponding matrix. This sum does not depend on the choice of the basis, and particularly we have $TrS = \sum \lambda_i$, where λ_i $(i = 1, 2, \ldots)$ are the eigenvalues of S. Covariance operators of Gaussian distributions on H have finite trace.

3.1.5

Theorem. *For every Gaussian distribution μ on H*

$$E_\mu e^{\lambda \|x\|^2} < +\infty$$

for any

$$\lambda < (2TrS)^{-1}.$$

This result, which is a particular case $(p = 2)$ of the theorem of Subsection 2.3.4, will be strengthened in the theorem of Subsection 3.1.10.

The integrability of the square of the norm with respect to an arbitrary Gaussian distribution on H has been previously proved by E. Mourier [3].

3.1.6

For $p = 2$ the topology $\mathfrak{T}_p$, given in Subsection 2.2.7, becomes *Sazonov's topology* in the space H. In noncoordinate form this topology is determined by the basis of neighborhoods of zero that consist of the sets $\mathcal{O}_S(0) = \{h : (Sh, h) < 1\}$, where S is an arbitrary S-operator on H. It is obvious that this topology is weaker than the metric topology of the space H.

Theorem (V. V. Sazonov). *In order that the functional χ be the characteristic functional of some distribution on H, it is necessary and sufficient that the following conditions be satisfied:*

a. *χ is positive-definite.*
b. $\chi(0) = 1$.
c. *χ is continuous in the $\mathfrak{T}_2$-topology.*

If these conditions are satisfied, then the corresponding distribution is determined uniquely.

PROOF. *Sufficiency.* Without loss of generality, we shall assume that the space H is represented in the form of l_2, and let χ_0 denote the restriction of

χ to $R_0^N \subset l_2$. From the conditions of the theorem the functional χ_0 satisfies all the assumptions of the theorem of Subsection 1.2.2. Thus there exists a unique distribution μ on R^N such that $\chi(h; \mu) = \chi(h)$, $h \in R_0^N$. We shall show that this distribution is concentrated on l_2. For this it is sufficient to show that

$$\lim_{r\to\infty} \lim_{n\to\infty} \mu\left\{ x : \sum_{j=1}^{n} x_j^2 \geqslant r^2 \right\} = 0. \tag{3.3}$$

Let $\epsilon > 0$ be arbitrary. Because of the continuity of χ in the $\mathfrak{T}_2$-topology, there exists an S-operator S_ϵ such that $|1 - \chi(h)| < \epsilon$ if $(S_\epsilon h, h) < 1$. If $(S_\epsilon h, h) \geqslant 1$ then it is obvious that $|1 - \chi(h)| \leqslant 2(S_\epsilon h, h)$ because $|\chi(h)| \leqslant 1$. Thus everywhere in R_0^N we have the inequality

$$|1 - \chi(h)| \leqslant \epsilon + 2(S_\epsilon h, h). \tag{3.4}$$

In order to estimate the measure of the event under the limits in condition (3.3), we shall use the inequality in the theorem of Subsection 1.2.10. Using inequality (3.4), after some elementary integrations, we obtain

$$\mu\left\{ x : \sum_{j=1}^{n} x_j^2 \geqslant r^2 \right\} \leqslant C\left(\epsilon + \frac{4}{r^2} \sum_{k=1}^{n} s_{\epsilon,k,k} \right), \tag{3.5}$$

where $s_{\epsilon,k,k}$ ($k = 1, 2, \ldots$) are the diagonal elements of the matrix of the operator S_ϵ in the natural basis in l_2.

Now, using inequality (3.5) and the fact that the trace of the operator S_ϵ is finite, we see that condition (3.3) is satisfied, and hence $\mu(l_2) = 1$ is established.

What remains is that the equality $\chi(h; \mu) = \chi(h)$ is true not only in R_0^N but also everywhere in l_2. This follows from the fact that R_0^N is everywhere dense in l_2 and the functionals $\chi(h)$ and $\chi(h; \mu)$ are both continuous in the l_2-norm, the first according to the assumption and the second according to the established fact that $\mu(l_2) = 1$.

The proof of the necessity part of the theorem is quite simple [16], hence we shall omit it. □

3.1.7

Returning to Gaussian distributions, consider the problem of the calculation of measures of balls in the space H. To do this we shall use the ordinary (one-dimensional) formula for inverting Fourier transforms. In principle, the method can be used in a more general case, but for simplicity we shall limit ourselves to balls, Gaussian measures, and Hilbert spaces.

Let $F(r) = \mu\{x : \|x\|^2 < r\}$, $r > 0$. It is obvious that F is the distribution function of the square of the norm, treated as a random variable in the space H with the probability measure μ. Further, let φ be the characteristic functional of this random variable. Then, of course,

$$\varphi(t) = \int_H e^{i\|x\|^2 t} \mu(dx) = \int_0^\infty e^{irt}\, dF(r), \tag{3.6}$$

and using the inversion formula (see, e.g., reference 13, p. 186), we obtain

$$F(r) = \frac{1}{2\pi} \int_{-\infty}^{+\infty} \frac{1 - e^{itr}}{it} \varphi(t)\, dt, \tag{3.7}$$

where the integral is taken in the principal value sense.

Assume for simplicity that the expectation of the Gaussian distribution μ is zero and denote by S the covariance operator of this distribution. Using as the basis the eigenelements $\{e_k\}$ of the operator S, we may write the equality

$$\|x\|^2 = \sum_{k=1}^\infty \lambda_k \gamma_k^2, \tag{3.8}$$

where $\lambda_1 \geqslant \lambda_2 \geqslant \cdots$ are the eigenvalues of the operator S in decreasing order (each written as many times as is its multiplicity), and $\gamma_1, \gamma_2, \ldots$ are independent normalized Gaussian random variables. In fact, we have

$$\|x\|^2 = \sum_{k=1}^\infty (x, e_k)^2, \qquad E_\mu(x, e_k) = 0,$$

$$E_\mu(x, e_k)^2 = \lambda_k,$$

and it is sufficient to put γ_k equal to $\lambda_k^{-1/2}(x, e_k)$ (without loss of generality we assume that there are no zero eigenvalues of the operator S). Substituting (3.8) into (3.6), taking independence into account, and integrating, we obtain

$$\varphi(t) = \prod_{k=1}^\infty (1 - 2it\lambda_k)^{-1/2}, \tag{3.9}$$

in which the infinite product is convergent because the trace of the S-operator S is finite (note again that the trace of an operator does not depend on the basis and is equal to the sum of all the eigenvalues).

Substituting (3.9) into (3.7), we obtain

$$F(r) = \frac{1}{2\pi} \int_{-\infty}^{+\infty} \frac{1 - e^{itr}}{it} \prod_{k=1}^\infty (1 - 2it\lambda_k)^{-1/2}\, dt, \tag{3.10}$$

which can be used as a basis for the estimation of Gaussian measures of balls in particular cases. Another possible method of estimation is connected with the application of the density of the distribution of the norm treated as a random variable. (See Subsection 3.1.9.)

In some particular cases, depending on the selection of the sequence $\{\lambda_k\}$, the infinite product in (3.9) can be expressed by elementary functions. In this way we can obtain a more convenient equation for estimating the measures of balls. One such case, where $\lambda_k = k^{-2}$, has been considered by G. E. Shilov and Fan Dick Tin [28]. In this case the explicit application of formula (3.10) gives the following value of the measure of a ball with its center at zero and the radius r.

$$F(r^2) = \frac{1}{2\pi} \int_{-\infty}^{+\infty} \frac{1 - e^{itr^2}}{it} \left(\frac{\pi\sqrt{2it}}{\sin \pi\sqrt{2it}} \right)^{1/2} dt.$$

3.1.8

Let us note that if $S^{1/2}$ means the positive square root of the covariance operator S of the Gaussian distribution μ on H, then the set $S^{1/2}H$ (the image of the space H by the transformation $S^{1/2}$) has μ-measure zero.[5] This fact can be deduced from the theorem of Subsection 3.1.4. Again using the basis of eigenelements of the operator S, we easily can give a simple explicit proof. Indeed, $x \in S^{1/2}H$ if and only if there exists an element $y \in H$ such that $S^{1/2}y = x$, that is, in the basis $\{e_k\}$ we have $x_k = \lambda_k^{1/2} y_k$. Therefore $x \in S^{1/2}H$ if and only if $\{x_k \lambda_k^{-1/2}\} \in l_2$. Thus we have to find the measure of the set of convergence of the series $\sum_{k=1}^{\infty} x_k^2/\lambda_k$. This series converges only on a set of measure zero, because at the points of convergence the general term should tend to zero and $\{x_k \lambda_k^{-1/2}\}$ cannot tend to zero on a set of positive measure, because $x_k \lambda_k^{-1/2}$ has for each k the standard Gaussian distribution. Note that in the proof we used the fact that the distribution μ is nondegenerate, by assuming that all the eigenvalues of the operator S are positive. But it is obvious in reality that we used the fact that the set of nonzero eigenvalues is infinite, therefore the fact that $\mu(S^{1/2}H) = 0$ is established under the assumption that there is no degeneracy on a finite-dimensional subspace.

3.1.9

In this subsection we shall estimate the density of the distribution of the square of the norm of a Gaussian random element in H. For simplicity we shall assume that the random element is defined on the probability space

---

[5] From this it obviously follows that the set SH also has measure zero.

$(H, \mathfrak{B}(H), \mu)$ as the identity mapping of H into itself and μ is the Gaussian distribution with expectation zero and the covariance operator S, whose eigenvalues are arranged in decreasing order $\lambda_1 \geqslant \lambda_2 \geqslant \ldots$, and each is written as many times as its multiplicity. Let κ denote the multiplicity of the maximal eigenvalue λ_1. Consider the two cases: $\kappa = 1$ and $\kappa > 1$ separately. First, let $\kappa > 1$. Using the basis of the eigenelements of the operator S, we obtain the formula (see (3.8))

$$\|x\|^2 = \lambda_1(\gamma_1^2 + \gamma_2^2 + \cdots + \gamma_\kappa^2) + \sum_{k=\kappa+1}^{\infty} \lambda_k \gamma_k^2. \tag{3.11}$$

Taking into account the total independence of the standard Gaussian random variables $\gamma_1, \gamma_2, \ldots,$ and denoting by ρ the density of the distribution of $\|x\|^2$, we obtain

$$\rho(u) = \int_0^u p_\kappa(u - v)\, d\Phi(v) \quad \text{for } u > 0,$$

$$\rho(u) = 0 \qquad\qquad \text{for } u \leqslant 0,$$

where p_κ is the density of the distribution of the sum $\lambda_1(\gamma_1^2 + \cdots + \gamma_\kappa^2)$, and Φ is the distribution function of the infinite sum of the remaining terms in (3.11). Substituting the expression for p_κ, which follows easily from the well-known (see, e.g., reference 42, p. 43) expression for the density of the chi-square distribution with κ degrees of freedom, we obtain

$$\rho(u) = \frac{u^{(\kappa-2)/2} e^{-u/2\lambda_1}}{2^{\kappa/2}\Gamma(\kappa/2)\lambda_1^{\kappa/2}} \int_0^u \left(1 - \frac{v}{u}\right)^{(\kappa-2)/2} e^{v/2\lambda_1}\, d\Phi(v) \tag{3.12}$$

If we now use the following formula (remembering that $\lambda_k \gamma_k^2 = (x, e_k)^2$)

$$\int_0^\infty \exp(v/2\lambda_1)\, d\Phi(v) = \int_H \exp\left\{ (1/2\lambda_1) \sum_{k=\kappa+1}^{\infty} \lambda_k \gamma_k^2 \right\} \mu(dx) \tag{3.13}$$

and (taking independence into account) perform the elementary integration, we obtain

$$\int_0^\infty e^{v/2\lambda_1}\, d\Phi(v) = \prod_{k=\kappa+1}^{\infty} \left(1 - \frac{\lambda_k}{\lambda_1}\right)^{-1/2}. \tag{3.14}$$

If we use this in the integral (3.12) replacing the function $(1 - v/u)^{(\kappa-2)/2}$

by one and extending the integration to the positive half line, we obtain the following estimate for the case $\kappa > 1$.

$$\rho(u) \leqslant \frac{u^{(\kappa-2)/2}e^{-u/2\lambda_1}}{2^{\kappa/2}\Gamma(\kappa/2)\lambda_1^{\kappa/2}} \prod_{k=\kappa+1}^{\infty}\left(1-\frac{\lambda_k}{\lambda_1}\right)^{-1/2}. \tag{3.15}$$

In the case $\kappa = 1$ we shall single out from the sum that expresses the square of the norm two summands rather than one, namely, $\lambda_1\gamma_1^2 + \lambda_2\gamma_2^2$. Let p denote the density of this sum (i.e., $\lambda_1\gamma_1^2 + \lambda_2\gamma_2^2$). Then,

$$p(u) = \frac{1}{2\pi(\lambda_1\lambda_2)^{1/2}} \int_0^u \frac{\exp\left\{-\dfrac{u-v}{2\lambda_1} - \dfrac{v}{2\lambda_2}\right\}}{[(u-v)v]^{1/2}}\,dv$$

$$\leqslant \frac{e^{-u/2\lambda_1}}{2\pi(\lambda_1\lambda_2)^{1/2}} \int_0^u \frac{dv}{[(u-v)v]^{1/2}} = \frac{e^{-u/2\lambda_1}}{2(\lambda_1\lambda_2)^{1/2}},$$

and if formulas (3.13) and (3.14) are now used (taking $\kappa = 2$), we obtain the following estimate of the density in the case $\kappa = 1$.

$$p(u) \leqslant \frac{e^{-u/2\lambda_1}}{2(\lambda_1\lambda_2)^{1/2}} \prod_{k=3}^{\infty}\left(1-\frac{\lambda_k}{\lambda_1}\right)^{-1/2}. \tag{3.16}$$

In principle, the relations obtained here can be used to calculate the Gaussian measures of balls. For example, in the case $\kappa > 1$ the integral in (3.12) can be written explicitly. But these relations, similar to formula (3.10) are of little use for explicit calculations. (In fact, there is no way to perform the calculations.) However, for the estimation of the values of the measures of balls these formulas might be useful.

As a corollary to the inequalities (3.15) and (3.16), note that the estimate $\mu\{x : \|x\| \geqslant r\} \leqslant Ke^{-r^2/2\sigma_\mu^2}$ from 2.3.4 may be strengthened.[6] Note also the following improvement of the theorem of Subsection 3.1.5: $E_\mu e^{\lambda\|x\|^2} < +\infty$, if $\lambda < 1/2\lambda_1$.

This result also follows from the theorem of Subsection 3.12, whose proof is surprisingly simple; namely, it is sufficient to write $\|x\|^2$ in the basis of eigenelements of the covariance operator and to carry out the cumbersome but elementary calculations.

[6]For example, for the particular case $\lambda_k = k^{-2}$ we obtain, as easily seen, the estimate

$$\mu\{x : \|x\| \geqslant r\} \leqslant \sqrt{6}\, e^{-r^2/2}.$$

3.1.10

Theorem. *Let λ_1 denote the maximal eigenvalue of the covariance operator S of the Gaussian distribution μ on the space H. Let $a_k = E_\mu(x, e_k)$ $(k = 1, 2, \ldots)$ be the coordinates of the expectation of the distribution μ in the basis of eigenelements of the operator S. Then,*

$$\int_H e^{\lambda \|x\|^2} \mu(dx) = \exp\left\{\lambda \sum_{k=1}^{\infty} \frac{a_k^2}{1 - 2\lambda\lambda_k}\right\} \prod_{k=1}^{\infty} (1 - 2\lambda\lambda_k)^{-1/2}$$

for each λ satisfying the condition $\lambda < 1/2\lambda_1$.

It should be noted that in the particular case $\lambda_k = k^{-2}$, $a_k = 0$ $(k = 1, 2, \ldots)$, we obtain

$$E_\mu e^{\lambda \|x\|^2} = \left(\frac{\pi\sqrt{2\lambda}}{\sin \pi\sqrt{2\lambda}}\right)^{1/2}, \qquad \lambda < 1/2;$$

and if, for example, $\lambda = \frac{1}{8}$, then

$$E_\mu e^{\|x\|^2/8} = (\pi/2)^{1/2}.$$

REMARK. It should be noted that the possibility of strengthening Theorem 2.3.3 in the particular case $p = 2$ is associated with the following two facts: (1) In a Hilbert space the matrix of the covariance operator can be assumed to be diagonal without loss of generality, because there exists a basis in which it is so, and the transformation into this basis is always made by a unitary operator, which does not change the norm of the element. (2) The form of the norm in the space l_2 is "compatible" with the form of the density of the Gaussian distribution. This is true not only for integration of functions of the norm but also for generalizations of different kinds.

3.1.11

REMARK. Let A be a bounded linear self-adjoint nonnegative operator in H. The quadratic form (Ax, x) is the square of the norm $(Ax, x) = \|A^{1/2}x\|^2$, where $A^{1/2}$ is the positive square root of the operator A. It is easy to see that if x is a Gaussian random element with expectation zero and covariance S, then $A^{1/2}x$ is also a Gaussian random element with expectation zero and covariance operator $A^{1/2}SA^{1/2}$. Therefore for the density of the distribution of the quadratic form (Ax, x) the estimates (3.15) and (3.16) are also true. (One should replace $\lambda_1, \lambda_2, \ldots$ by the eigenvalues of the operator $A^{1/2}SA^{1/2}$).

For the quadratic form (Ax,x) the theorem analogous to the theorem of Subsection 3.1.10 is also true. In particular, we have

$$E_\mu e^{\lambda(Ax,x)} < +\infty \qquad \text{for any } \lambda < 1/2\lambda_1,$$

where λ_1 is the maximal eigenvalue of the operator $A^{1/2}SA^{1/2}$.

3.1.12

Lemma. *Let A and B be bounded linear operators on H and let $A \geqslant 0$. The sets of nonzero eigenvalues of the operators AB, BA, and $A^{1/2}BA^{1/2}$ are identical.*

PROOF. Let $ABh = \lambda h$, $h \neq 0$, $\lambda \neq 0$. If $g = A^{1/2}Bh$, then $A^{1/2}BA^{1/2}g = A^{1/2}BABh = \lambda A^{1/2}Bh = \lambda g$ and $g \neq 0$ because $A^{1/2}g = \lambda h$. Conversely, if $A^{1/2}BA^{1/2}h = \lambda h$, then putting $g = A^{1/2}h$, we have $ABg = A^{1/2}A^{1/2}Bg = A^{1/2}A^{1/2}BA^{1/2}h = \lambda A^{1/2}h = \lambda g$ and $A^{1/2}Bg = \lambda h$. Thus the sets of nonzero eigenvalues of the operators AB and $A^{1/2}BA^{1/2}$ are identical. The identity of the sets of nonzero eigenvalues of the operators BA and $A^{1/2}BA^{1/2}$ can be shown similarly. □

3.2 The Estimation of the Convergence Rate in the Central Limit Theorem

3.2.1

Let $x_1, x_2, \ldots$ be a sequence of independent, identically distributed random elements in H. We shall assume for simplicity that the expectation of x_1 is zero and denote by S its covariance operator, for which we assume that its trace is finite ($TrS < +\infty$). This condition is obviously equivalent to the condition $E\|x_1\|^2 < +\infty$ (recall that E means expectation). Without this condition the central limit theorem does not hold.

Thus we have by assumption

$$E(h,x_1) = 0, \qquad E(h,x_1)^2 = (Sh,h), \quad h \in H. \tag{3.17}$$

Further, let y_n denote the normalized sum

$$y_n = n^{-1/2}(x_1 + x_2 = \cdots = x_n). \tag{3.18}$$

From (3.17) it follows that for each n the random element y_n has expectation zero and covariance operator S. Let x denote the Gaussian random element in H with expectation zero and the same covariance operator S (the existence of such a Gaussian element follows from the theorem of Subsection 3.1.4). According to the theorem of Subsection 2.3.2, the distributions of normalized sums, (3.18) converge to the distribution of the Gaussian random element x. Our aim is to estimate the rate of this convergence.

3.2.2

First, we shall consider separately the finite-dimensional case $H = R^k$. In this case the approximation is much better than in the general (infinite-dimensional) case; however, we make the very limiting assumption of the independence of the coordinates of the random vectors in the cumulative sums. Although the noncorrelation of the coordinates can always be achieved by a corresponding choice of base, the independence requirement obviously limits the generality.

Let

$$\Delta_n(k) = \sup_{a \in R^k, r>0} \left| P\{\omega : \|y_n - a\| \leqslant r\} - p\{\omega : \|x - a\| \leqslant r\} \right|.$$

Theorem. *The following inequality holds.*

$$\Delta_n(k) \leqslant \frac{2c}{n^{1/2}} \sum_{j=1}^{k} \frac{E|x_{1,j}|^3}{\left(E|x_{1,j}|^2\right)^{3/2}}, \tag{3.19}$$

where c is an absolute constant that also occurs in the one-dimensional Berry–Esseen's inequality (for distribution functions).

PROOF. Note that $\|y_n - a\| \leqslant r$ means that

$$\sum_{j=1}^{k} \left[n^{-1/2} \sum_{i=1}^{n} x_{i,j} - a_j \right]^2 \leqslant r^2.$$

Let F and $\tilde{F}$ be the distribution functions of the following random variables, respectively.

$$\sum_{j=1}^{k-1} \left[n^{-1/2} \sum_{i=1}^{n} x_{i,j} - a_j \right]^2, \quad \left[n^{-1/2} \sum_{i=1}^{n} x_{i,j} - a_k \right]^2.$$

Using the mutual independence of the coordinates of the independent vectors x_i, we obtain

$$P\{\omega : \|y_n(\omega) - a\| \leqslant r\} = \int_{-\infty}^{+\infty} F(r^2 - t)\, d\tilde{F}(t). \tag{3.20}$$

Similarly, one can also obtain the expression

$$P\{\omega : \|x(\omega) - a\| \leqslant r\} = \int_{-\infty}^{+\infty} \Phi(r^2 - t)\, d\tilde{\Phi}(t), \qquad (3.21)$$

where Φ and $\tilde{\Phi}$ denote the distribution functions of the corresponding random variables

$$\sum_{j=1}^{k-1} (x_j - a_j)^2 \qquad \text{and} \qquad (x_k - a_k)^2.$$

Taking the difference of (3.21) and (3.20) after transformations, we get

$$\Delta_n(k) \leqslant \Delta_n(k-1) + \sup_{t,a} |\tilde{F}(t) - \tilde{\Phi}(t)|.$$

The second summand on the right-hand side does not exceed the value

$$2 \sup_t \left| P\left\{ \omega : n^{-1/2} \sum_{i=1}^{n} x_{i,k} < t \right\} - \left(2\pi\sigma_k^2\right)^{-1/2} \int_{-\infty}^{t} e^{-u^2/2\sigma_k^2}\, du \right|,$$

where $\sigma_k^2 = Ex_{1,k}^2$. Using the Berry-Esseen estimate (see, e.g., reference 13, p. 288), we have

$$\Delta_n(k) \leqslant \Delta_n(k-1) + \frac{2c}{n^{1/2}} \frac{E|x_{1,k}|^3}{\left(E|x_{1,k}|^2\right)^{3/2}}.$$

When we apply the same reasoning to $\Delta_n(k-1), \Delta_n(k-2), \ldots, \Delta_n(1)$, the proof of the theorem is complete. □

REMARK. The inequality (3.19) remains valid if the expectation of x_1 is not zero. One can also consider the case when the random elements may not be identically distributed. Then, instead of inequality (3.19), one has

$$\Delta_n(k) \leqslant \sum_{j=1}^{k} L_n(j),$$

where

$$L_n(j) = \frac{\sum_{i=1}^{n} E|x_{i,j}|^3}{\left(\sum_{i=1}^{n} E|x_{i,j}|^2\right)^{3/2}}$$

is *Liapunov's ratio* for the jth coordinate. In this case the meaning of $\Delta_n(k)$ should be changed correspondingly. This is obvious because now, instead of division by $n^{1/2}$, one has to apply the linear diagonal operator to the sum of n random elements. The action of this operator on a given element in R^k is division of the jth coordinate ($j = 1, 2, \ldots, k$) by $(\sum_{i=1}^{n} E|x_{i,j}|^2)^{1/2}$. Accordingly, for the vector x one should take the standard Gaussian vector (which has the identity operator as the covariance operator).

3.2.3

We proceed now to the infinite-dimensional case. Let A be a nonnegative bounded linear self-adjoint operator in H and let

$$\Delta_n(A, S, x_1) = \sup |P\{(Ay_n, y_n) < r\} - P\{(Ax, x) < r\}|. \tag{3.22}$$

We want to estimate the rate of decrease of Δ_n. The method is based on the use of the theorem of Subsection 3.1.4. We shall first prove some lemmas.

Let χ_n and χ be the characteristic functionals of the normalized sum y_n and the Gaussian random element x, respectively. Let

$$L = \sup_{\|h\|=1} \frac{E|(h, x_1)|^3}{(Sh, h)}, \qquad h \in H. \tag{3.23}$$

Lemma. *For* $\|h\| < n^{1/2}/5L$ *we have the inequality*

$$|\chi_n(h) - \chi(h)| \leqslant 14L\|h\|/(3en^{1/2}). \tag{3.24}$$

PROOF. The proof is based on the known (see, e.g., reference 43, p. 112) estimate of the distance from $e^{-t^2/2}$ to the characteristic functional of normalized and centered sums of independent (numerical) random variables. In order to be able to use this inequality put $(Sh, h)^{1/2} = t$, and using the theorem of Subsection 3.1.4 let us write $\chi(h)$ in the form $e^{-t^2/2}$. Further, we have

$$\chi_n(h) = Ee^{i(y_n, h)} = E\exp\left\{ it\frac{\sum_{k=1}^{n}(x_k, h)}{n^{1/2}(Sh, h)^{1/2}} \right\};$$

and it is obvious that for each fixed $h \in H$, we can look at the complex number $\chi_n(h)$ as the value of the characteristic functional of the normalized and centered (see relations (3.17)) sum of independent and identically distributed real random variables (x_k,h) at the fixed point $t=(Sh,h)^{1/2}$. For the proof of the lemma it remains only to use the aforementioned estimate for the numerical case and the elementary inequality $te^{-t} \leqslant e^{-1}$, $t>0$. □

REMARK. Inequality (3.24) is nontrivial, of course, only in the case in which the number L determined in (3.13) is finite. If the Hilbert space H is finite dimensional, then the condition $L < +\infty$ follows easily from the existence of the third moment of the norm of the random element x_1. In the infinite-dimensional case the condition $L < +\infty$ does not follow from the existence of moments. We shall give an example in which moments of all orders exist, but where $L = +\infty$. But first, let us note that the condition $L < +\infty$ will be satisfied if the random element x_1 (and therefore all x_ks) is bounded with probability 1; that is, if

$$P\{\omega : \|x_1\| \leqslant L_0\} = 1 \qquad \text{for some } L_0 > 0. \tag{3.25}$$

In this case obviously $L \leqslant L_0$. We note that in the finite-dimensional case the fact that L is finite is equivalent to the fact that the upper limit of the Liapunov ratios

$$L_{\max} = \sup_h \frac{E|(x_1,h)|^3}{(Sh,h)^{3/2}}$$

is finite. In the infinite-dimensional case, from the condition $L_{\max} < +\infty$ it follows as before, that $L < +\infty$, but the converse is not true (Condition (3.25) from which the finiteness of L follows does not imply that $L_{\max}$ is finite). Therefore it is better to have an estimate with the constant L than with the constant $L_{\max}$.

We shall now give our example.

EXAMPLE. Let the basic probability space be the interval $(0,1)$ with Lebesgue measure and let the random element x_1 be defined by

$$x_1(\omega) = \sum_{k=1}^{\infty} x_{1,k}(\omega)e_k,$$

where $\{e_k\}$ is some basis in H and

$$x_{1,k}(\omega) = a_k \qquad \text{if } \omega \in (0,\omega_k),$$
$$x_{1,k}(\omega) = 0 \qquad \text{if } \omega \in [\omega_k, 1).$$

It is easy to show that for an arbitrary sequence of numbers $a_k \to +\infty$, $L = +\infty$; but for properly chosen sequences $\{a_k\}$, $(a_k \to +\infty)$, and $\{\omega_k\}$, $E\|x_1\|^t < +\infty$ for all $t \geqslant 0$ (e.g., take $a_k \sim k$ and $\omega_k \sim e^{-k}$).

3.2.4

Lemma. *Let $A: H \to H$ be a nonnegative linear self-adjoint operator with a finite trace and let λ be a positive number. Then, the following estimate holds.*

$$|Ee^{-\lambda(Ay_n, y_n)} - Ee^{-\lambda(Ax,x)}| \leqslant C(2\lambda/n)^{1/2},$$

where

$$C \leqslant \left(\frac{14}{3e} + 10\right)(TrA)^{1/2}L.$$

PROOF. According to the theorem of Subsection 3.1.4, one can treat the expression $e^{-\lambda(Ay_n, y_n)}$ as the value of the characteristic functional of the Gaussian distribution $\mu = \mu_{2\lambda A}$ on H with zero mean and covariance operator $2\lambda A$ (recall that according to the assumption, $2\lambda A$ is an S-operator) at the "random point" y_n. Therefore

$$e^{-\lambda(Ay_n, y_n)} = \int_H e^{i(y_n, h)}\mu(dh), \quad h \in H.$$

A similar formula can be written for the expression $e^{-\lambda(Ax,x)}$. Writing the corresponding difference and changing the order of integration, we obtain

$$\begin{aligned}
&|Ee^{-\lambda(Ay_n, y_n)} - Ee^{-\lambda(Ax,x)}| \\
&\quad \leqslant \int_H |Ee^{i(y_n,h)} - Ee^{i(x,h)}|\,\mu\,dh \\
&\quad = \int_H |\chi_n(h) - \chi(h)|\,\mu\,dh.
\end{aligned}$$

Let us write the integral on the right-hand side of this inequality as the sum of two integrals, one over the ball $\|h\| \leqslant n^{1/2}/5L$, the other over the complement of this ball. In the first of these integrals replace the difference $|\chi_n(h) - \chi(h)|$ by estimate (3.24) of the lemma of Subsection 3.2.3; whereas in the second integral replace this difference by 2 and estimate the measure of the complement of the ball using Chebyshev's inequality. Then, the estimate of the difference under consideration is given by the number

$$\frac{L}{n^{1/2}}\left(\frac{14}{3e} + 10\right)\int_H \|h\|\,\mu(dh);$$

and it is sufficient to apply Hölder's inequality and the formula

$$\int_H \|h\|^2 \mu\, dh = Tr(2\lambda A) = 2\lambda Tr A,$$

which can easily be obtained by writing $\|h\|^2$ in the basis of eigenelements of the operator A. □

3.2.5

Let us go to the next lemma. Write the quadratic form (Ax, x) as $\|A^{1/2}x\|^2$, where $A^{1/2}$ is the positive square root of the operator A. As noticed in Subsection 3.1.11, $A^{1/2}x$ is also a Gaussian random element, with mean zero and covariance operator $A^{1/2}SA^{1/2}$. Let $\lambda_1 \geqslant \lambda_2 \geqslant \cdots$ be the eigenvalues of this operator written in decreasing order and their multiplicity taken into account. Let κ be the multiplicity of λ_1. The values of the constants in the final estimates of the convergence rate depend on κ. Although these constants are not the best possible ones, we shall not attempt to improve them here. Because there is no point in considering all the cases; we shall take for simplicity $\kappa \leqslant 2$. Put

$$\eta = \exp\left\{ -(2\lambda_1)^{-1}(Ax, x)\right\} \tag{3.26}$$

and let F be the distribution function of the random variable η.

Lemma. *The following estimate of the density is valid.*

$$\sup_u F'(u) \leqslant \left(\frac{\lambda_1}{\lambda_2}\right)^{1/2} \prod_{k=3}^{\infty} \left(1 - \frac{\lambda_k}{\lambda_1}\right)^{-1/2}.$$

PROOF. We have $0 < \eta < 1$ with probability 1. Therefore $F'(v) = 0$ if $v \notin [0, 1]$; and inside this interval (3.26) implies the relation

$$F'(v) = v^{-1} 2\lambda_1 \rho(-2\lambda_1 \ln v),$$

where ρ is the probability density function of (Ax, x). Therefore denoting $-2\lambda_1 \ln v$ by u, we obtain

$$\sup_u F'(u) = 2\lambda_1 \sup_u \left\{ e^{u/2\lambda_1} \rho(u)\right\},$$

and only the remark of Subsection 3.1.11 is needed to complete the proof of the lemma. □

3.2.6

Let us now consider the expression $\Delta_n = \Delta_n(A, S, x_1)$ given by (3.22). First, let us assume that the nonnegative operator A has a finite trace. The estimate for this case is given by the following theorem.

Theorem. *The following inequality holds.*

$$\Delta_n(A, S, x_1) \leqslant \frac{C_1(A, S)}{1-2\epsilon}\,\frac{1}{\ln n} + \frac{C_2(A, S, x_1)}{n^{\epsilon}},$$
$$(n = 1, 2, \ldots),$$

where $0 < \epsilon < 1/2$ *is arbitrary and the constants* C_1 *and* C_2 *can be taken as*[7]

$$C_1(A, S) = C_1(AS) = \frac{48}{\pi}\left(\frac{\lambda_1}{\lambda_2}\right)^{1/2} \prod_{k=3}^{\infty}\left(1 - \frac{\lambda_\kappa}{\lambda_1}\right)^{-1/2},$$
$$C_2(A, S, x_1) = \frac{2}{\pi}\left(\frac{14}{3e} + 10\right) L(TrA/\lambda_1)^{1/2}.$$

PROOF. Let F_n be the distribution function of the random variable

$$\eta_n = \exp\left\{-\frac{1}{2\lambda_1}(Ay_n, y_n)\right\},$$

and F the distribution function of the random variable η. Note that the random variable Δ_n may be written as

$$\Delta_n = \sup_u |F_n(u) - F(u)|, \tag{3.27}$$

and let us write the Berry–Esseen inequality (see, e.g., reference 42, p. 161) as follows.

$$\sup_u |F_n(u) - F(u)| \leqslant (2/\pi)\int_0^u |\varphi_n(t) - \varphi(t)|t^{-1}\,dt + 24(\pi u)^{-1}\sup_u F'(u). \tag{3.28}$$

Here φ_n and φ are the characteristic functions of the distributions F_n and F, respectively. To estimate the second summand on the right-hand side of inequality (3.28) we proved the lemma of Subsection 3.2.5. Therefore we

[7] Recall that we consider the case $\kappa \leqslant 2$. For $\kappa > 2$ the theorem remains true, but the values of the constants C_1 and C_2 will be changed.

shall now concentrate on the first summand. After elementary considerations we obtain the inequality

$$|\varphi_n(t) - \varphi(t)| \leq \sum_{k=1}^{\infty} (k!)^{-1} t^k \left| E(\eta_n^k - \eta^k) \right|, \quad t > 0.$$

Now each summand on the right-hand side of this inequality will be estimated using the lemma of Subsection 3.2.4 for $\lambda = k/2\lambda_1$, $k = 1, 2, \ldots$. Thus we obtain the inequality

$$|\varphi_n(t) - \varphi(t)| \leq C(\lambda_1 n)^{-1/2} t e^t,$$

which according to relations (3.27) and (3.28) gives the estimate

$$\Delta_n \leq 2C\pi^{-1}(\lambda_1 n)^{-1/2} e^{v} + 24\pi^{-1} v^{-1} \sup_u F'(u).$$

Choosing $v = [(1/2) - \epsilon] \ln n$, $0 < \epsilon < 1/2$, using the lemma of Subsection 3.2.5 and the inequality for the constant C from the lemma of Subsection 3.2.4, we obtain the required estimate. □

3.2.7

We now want to omit the condition that the trace of the operator A be finite and to consider the general case. We note readily that the investigation of the general case of an arbitrary nonnegative operator A can be reduced to the investigation of the identity operator $A = I$. We shall consider this case only and estimate the difference $\Delta_n(I, S, x_1)$. Let e_1, $e_2, \ldots$ be a complete orthonormal system of eigenelements of the operator S and let $\lambda_1 \geq \lambda_2 \geq \cdots$ be the corresponding eigenvalues, written in decreasing order with their multiplicity taken into account. First, we suppose that the operator $S^{1/2}$ also has a finite trace, that is, the following condition is satisfied.

$$\sum_{k=1}^{\infty} \lambda_k^{1/2} < +\infty \tag{3.29}$$

Let $S^{-\alpha}(\alpha > 0)$ denote the operator, whose action on the element $h \in H$ means multiplication of the coordinate (h, e_k) by the number $\lambda_k^{-\alpha}$. The operator $S^{-\alpha}$ is not defined and bounded everywhere. But by assumption (3.29) the elements $\tilde{y}_n = S^{-1/4} y_n$ and $\tilde{x} = S^{-1/4} x$ exist with probability 1, have mean zero and covariance operator $S^{1/2}$, and $\tilde{x}$ is Gaussian. Taking this into account and writing the balls $\|y_n\|^2 \leq r$ and $\|x\|^2 \leq r$ as ellipsoids $(S^{1/2}\tilde{y}_n, \tilde{y}_n) \leq r$ and $(S^{1/2}\tilde{x}, \tilde{x}) \leq r$, respectively, we obtain the relations

$$\Delta_n(I, S, x_1) = \Delta_n(S^{1/2}, S^{1/2}, S^{-1/4} x_1).$$

Application of the theorem of Subsection 3.2.6 now gives the following result.

Theorem. *The following inequality holds.*

$$\Delta_n(I,S,x_1) \leqslant \frac{C_1(I,S)}{1-2\epsilon}\,\frac{1}{\ln n} + \frac{C_2(I,S,S^{-1/4}x_1)}{n^{\epsilon}},$$
$$(n = 1,2,\dots).$$

Here $0<\epsilon<1/2$ is arbitrary, and for estimation of the numbers $C_1(I,S)$ and $C_2(I,S,S^{-1/4}x_1)$ one can use the expressions given in the theorem of Subsection 3.2.6.

REMARK. If condition (3.29) is not satisfied, then not all the previous considerations are valid. But using the method of finite-dimensional approximations and the equality

$$\Delta_n(I,S,x_1) = \Delta_n(S,I,S^{-1/2}x_1),$$

which does not need an explanation in the finite-dimensional case, we can easily find an estimate of the difference $\Delta_n(I,S,x_1)$ without using condition (3.29). The result will not differ from the theorem of Subsection 3.2.7, for $L = L(S^{-1/2}x_1)$ one has to use the limit of the corresponding expressions for finite-dimensional estimates.

Therefore the theorem of Subsection 3.2.6 formally remains true for the general case of an arbitrary nonnegative operator A (without assuming that the trace is finite). But one should recognize the difference between the two estimates obtained in the case $TrA < +\infty$ (theorem of Subsection 3.2.6) and without this condition (theorem of Subsection 3.2.7). The only, but important, difference is that in the general case we cannot use the simple sufficient condition for the constant $L(S^{-1/4}x_1)$ (or for $L(S^{-1/2}x_1)$ in the case where condition (3.29) is not satisfied) to be finite. For example, the condition that x_1 is bounded with probability 1, which implies that the constant $L(x_1)$ in the theorem of Subsection 3.2.6 is finite, does not imply that the theorem of Subsection 3.2.7 is nontrivial because $L(S^{-1/4}x_1)$ can be infinite.[8] Therefore it is more convenient for the case of convergence on balls to have an estimate with the constant $L(x_1)$ instead of the constant $L(S^{-1/4}x_1)$. Such an estimate can be obtained (still of order 0 $(\ln^{-1} n)$) by

---

[8]Here is a simple example. Let $\xi_1, \xi_2, \dots$ be a sequence of independent random variables with distributions given by $P\{\xi_k = \pm a_k\} = \lambda_k/2$ and $P\{\xi_k = 0\} = 1-\lambda_k$, $(k = 1,2,\dots)$. Then, $E\xi_k = 0$ and $E\xi_k^2 = \lambda_k a_k^2$. Assume that $\sum a_k^2 < +\infty$, $a_k^4/\lambda_k \to +\infty$. It is easy to see that $x = \{\xi_k\} \epsilon l_2$ with probability 1. A simple computation shows that in this case $L(S^{-1/4}x_1) = +\infty$. Boundedness with probability 1 is obvious.

using the truncation method (finite-dimensional approximations) and the theorem of Subsection 3.2.6 for the estimate in the finite-dimensional case. In this case TrS will be equal to the dimension of the space, hence the constant in the finite-dimensional estimate increases as $k^{1/2}$, where k is the dimension. Such a slow increase of the constant for $k \to \infty$ allows us to find the estimate for the infinite-dimensional case by some slight restrictions on the convergence rate of the series of eigenvalues (the logarithmic decrease of the remainder of this series suffices).

Finally, we shall make one more remark. All our estimates are obtained for ellipsoids and balls with a fixed center at zero. If arbitrary centers are taken (which can vary through the whole space), then the estimates will not be uniform with respect to the centers (the constant C_1 will increase if the center tends to infinity, and the constant C_2 will not depend on the center). The estimates will be uniform, however, if the centers vary inside some fixed ball.

3.3 Some Other Problems

3.3.1

In this subsection we shall consider three main problems; one is to estimate large deviations of sums of independent random elements in a Hilbert space. Let $x_1, x_2, \ldots$ be independent centered random elements in H and let $y_n = n^{-1/2}\sum_{i=1}^{n} x_i$ be their normalized sums. The term "normalized sum" will be taken conditionally, because we do not require the random elements that constitute these sums to be identically distributed. We do not even require that their covariance operators be identical. We only assume that their expectations are zero and that

$$P\{\omega : \|x_i(\omega)\| \leqslant L\} = 1, \qquad (i = 1, 2, \ldots); \tag{3.30}$$

that is, they are bounded in norm with probability 1 by a common constant $L > 0$.

Let A be some nonnegative bounded linear self-adjoint operator in H. First, we assume that A has finite trace. We would like to estimate the probability of the event $\{(Ay_n, y_n) \geqslant r^2\}$ for large r. To do this we use inequality (2.21) from Subsection 2.3.3, the known elementary estimate

$$Ee^{\xi} \leqslant e^{L^2}, \tag{3.31}$$

where ξ is a centered random variable whose absolute value is bounded by the number L with probability 1, and the relation

$$\exp\left\{\frac{1}{2}(Af, f)\right\} = \int_H e^{(-f,h)}\mu(dh), \qquad f \in H, \tag{3.32}$$

in which μ is the Gaussian distribution with mean zero and covariance operator A. The existence of μ follows from the theorem of Subsection 3.1.4. Formula (3.32) can be obtained in different ways, the simplest is explicit integration using the basis of eigenelements of the operator A for the inner product (f, h). Using such a basis, we obtain

$$\int_H \prod_{k=1}^{\infty} e^{-(f_k, h_k)} \mu(dh) = \prod_{k=1}^{\infty} (2\pi\lambda_k)^{-1/2} \int_{-\infty}^{\infty} \exp\{-f_k u - (u^2/2\lambda_k)\} du$$
$$= \prod_{k=1}^{\infty} \exp\{\lambda_k f_k^2/2\},$$

which establishes (3.32).

Theorem. *The following inequality holds.*

$$P\{(Ay_n, y_n) \geqslant r^2\} \leqslant er^2(4L^2TrA)^{-1} \exp\{-r^2/(4L^2TrA)\},$$
$$r^2 \geqslant 4L^2TrA.$$

PROOF. Take a positive parameter t^2 and write the condition $(Ay_n, y_n) \geqslant r^2$ in the form

$$\exp\{(Ay_n, y_n)t^2/2\} \geqslant \exp\{t^2r^2/2\}.$$

First, we apply Chebyshev's inequality, then use the relation (3.32) at the "random points" $f = ty_n$, and finally, the estimate (3.31). Then, taking into account condition (3.30), and the independence of the random elements $x_1, x_2, \ldots,$ we obtain

$$P\{(Ay_n, y_n) \geqslant r^2\} \leqslant e^{-t^2r^2/2} \int_H Ee^{-(ty_n, h)} \mu(dh)$$
$$= e^{-t^2r^2/2} \int_H \prod_{i=1}^{n} Ee^{-t(x_i, h)/\sqrt{n}} \mu(dh)$$
$$\leqslant e^{-t^2r^2/2} \int_H e^{t^2L^2\|h\|^2} \mu(dh).$$

Expanding the exponential function under the integral and using inequality (2.21) from 2.3.3, we obtain

$$P\{(Ay_n, y_n) \geqslant r^2\} \leqslant e^{-t^2r^2/2} \left[1 + \sum_{k=1}^{\infty} \frac{(2t^2L^2TrA)^k}{k!\sqrt{\pi}} \Gamma((2k+1)/2)\right].$$

Imposing the condition $2t^2L^2TrA < 1$ and applying the obvious inequality[9]

$$\Gamma(k+\tfrac{1}{2}) \leqslant \sqrt{\pi}\, k! \qquad (k=1,2,\dots),$$

we obtain the inequality

$$P\{(Ay_n, y_n) \geqslant r^2\} \leqslant e^{-t^2r^2/2}(1-2t^2L^2TrA)^{-1}.$$

Now, all that remains is finding the minimum with respect to t^2 of the expression on the right-hand side of the last inequality. We do this by differentiation, and hence we obtain the desired inequality (the condition $r^2 \geqslant 4L^2TrA$ is used to assure that the minimum point is nonnegative). □

REMARK. We now assume that all random elements x_i have the same covariance operator S and that S satisfies condition (3.29) from Subsection 3.2.7. Further, let $S^{-\alpha}(\alpha>0)$ have the same meaning as in Subsection 3.2.7. The obvious identity

$$(Ay_n, y_n) = (S^{1/4}AS^{1/4}S^{-1/4}y_n, S^{-1/4}y_n),$$

where $S^{1/4}$ is the positive square root of $S^{1/2}$, shows that under these assumptions the preceding inequality also holds without the condition $TrA < +\infty$, provided that L is a constant which bounds the random element $S^{-1/4}x_1$ with probability 1, and TrA is replaced by $Tr(S^{1/4}AS^{1/4})$. Note that according to the lemma of Subsection 3.1.12, $Tr(S^{1/4}AS^{1/4}) = Tr(AS^{1/2}) < +\infty$ because according to the assumption we have $TrS^{1/2} < +\infty$ and A is a bounded operator.

3.3.2

Let x_1 and x_2 denote Gaussian random elements with values in H. The inner product (x_1, x_2) will be an ordinary (real) random variable. Our problem is now to investigate the distribution of this random variable. We shall show that this problem can be reduced to an investigation of the distribution of some quadratic functional of a Gaussian random element in the Cartesian product $H \times H$.

The elements of the product $H \times H$ are pairs

$$u = \begin{pmatrix} u_1 \\ u_2 \end{pmatrix}, \qquad u_i \in H, \quad i=1,2,$$

[9] We increase the constant in this estimate (we may at least remove the factor $\sqrt{\pi}$) in order to simplify the final result. A similar remark should be made about applying (3.31).

The set of these pairs becomes the real separable Hilbert space with the usual linear operations and the inner product given by

$$\langle u,v\rangle=(u_1,v_1)+(u_2,v_2). \tag{3.33}$$

The operators that transform $H\times H$ into itself, can be written as matrices whose elements are operators that transform H into itself. It is easy to show that the formal rules of composition of operators, and their operations on elements, are similar to the corresponding rules of the vector calculus of the second order; that is,

$$\begin{pmatrix}A_{11}, & A_{12}\\ A_{21}, & A_{22}\end{pmatrix}\begin{pmatrix}B_{11}, & B_{12}\\ B_{21}, & B_{22}\end{pmatrix}=\begin{pmatrix}A_{11}B_{11}+A_{12}B_{21}, & A_{11}B_{12}+A_{12}B_{22}\\ A_{21}B_{11}+A_{22}B_{21}, & A_{21}B_{12}+A_{22}B_{22}\end{pmatrix} \tag{3.34}$$

$$\begin{pmatrix}A_{11}, & A_{12}\\ A_{21}, & A_{22}\end{pmatrix}\begin{pmatrix}u_1\\ u_2\end{pmatrix}=\begin{pmatrix}A_{11}u_1+A_{12}u_2\\ A_{21}u_1+A_{22}u_2\end{pmatrix}. \tag{3.35}$$

Returning to Gaussian random elements x_i $(i=1,2)$, assume for simplicity, that their means are zero and let S_i denote the covariance operators, and T their cross covariance. These operators are determined by the relations

$$(S_ih,h)=E(h,x_i)^2, \qquad (Th,h)=E(h,x_1)(h,x_2). \tag{3.36}$$

Let us consider the random element

$$x=\begin{pmatrix}x_1\\ x_2\end{pmatrix}$$

in the Hilbert space $H\times H$.

This element does not necessarily have a Gaussian distribution (even for $H=R^1$, see Subsection 2.1.7). But we shall make the assumption that x is a Gaussian element (this will be true, for example, if x_1 and x_2 are independent). Using equations (3.33), (3.35), and (3.36), and making a simple transformation, we can easily show that the covariance operator of an element x has the form

$$K=\begin{pmatrix}S_1, & T^*\\ T, & S_2\end{pmatrix}, \tag{3.37}$$

where T^* is the operator adjoint to T.

Let A denote the operator for which $A_{11}=A_{22}=0$, $A_{12}=A_{21}=\frac{1}{2}I$, where 0 and I are, respectively, the zero and the identity operator on H. Consider the quadratic form $\langle Ax,x\rangle$. Using equalities (3.33) and (3.35), we see that this quadratic form is identical with the inner product (x_1,x_2). Thus applying the theorem about the distribution of the quadratic form of Gaussian random elements in the Hilbert space, proved by I. A. Ibragimov [44],[10] we obtain the following result.

Theorem. *The inner product* (x_1, x_2) *has the same distribution as the sum*

$$\sum_{k=1}^{\infty} \lambda_k \gamma_k^2,$$

where γ_k *are independent Gaussian random variables with expectations zero and variance one, and the* λ_k *are the eigenvalues of the linear operator* KA.

REMARK. According to the lemma of Subsection 3.1.12, the set of nonzero eigenvalues of a completely continuous, but not necessarily self-adjoint operator KA, is identical with an at most countable set of eigenvalues of the self-adjoint completely continuous operator $K^{1/2}AK^{1/2}$, which are concentrated on the real line.

3.3.3

From formulas (3.34) and (3.37) we see that the λ_k are the values of λ for which the following system of operator equations has a nontrivial solution.

$$\begin{aligned} T^* u_1 + S_1 u_2 &= 2\lambda u_1 \\ S_2 u_1 + T u_2 &= 2\lambda u_2, \end{aligned} \qquad \begin{pmatrix} u_1 \\ u_2 \end{pmatrix} \in H \times H. \tag{3.38}$$

Finding the eigenvalues of the system (3.38) may be simplified in some particular cases. Consider the two most important cases: (1) $S_1 = S_2$, $T = T^*$ and (2) $T = 0$. The eigenvalues of the system (3.38), for these two cases are described in the following two lemmas.

Lemma. *The set of eigenvalues of the system* (3.38) *for* $S_1 = S_2$ *and* $T = T^*$ *is the union of the sets of eigenvalues of the operators* $1/2(T + S_1)$ *and* $1/2(T - S_1)$.

PROOF. Let λ be an eigenvalue of the system (3.38), and let u_1, u_2 be the coordinates of the corresponding nontrivial solution. Then either $u_1 + u_2 \neq 0$ or $u_1 - u_2 \neq 0$. It is easy to see that in the first case λ will be an eigenvalue of the operator $1/2(T + S_1)$, and in the second, of the operator $1/2(T - S_1)$. The converse can be verified similarly. □

[10]The theorem in reference 44 is proved with the additional assumption of complete continuity of the operator of the quadratic form. But a slight change of the author's considerations enables us to remove this assumption.

3.3.4

Lemma. *The set of nonzero eigenvalues of the system,* (3.38), *for the case* $T=0$ (*i.e., for independent* x_1 *and* x_2) *is identical with the set of numbers of the form* $\neq \nu^{1/2}/2$, *where* ν *is a nonzero eigenvalue of the operator* S_1S_2 (*and therefore according to the lemma of Subsection* 3.1.12, $\nu > 0$).

PROOF. Let $S_1u_2 = 2\lambda u_1$, $S_2u_1 = 2\lambda u_2$, and $\lambda \neq 0$. Then $S_1S_2u_1 = 2\lambda S_1u_2 = 4\lambda^2 u_1$. Conversely, let $S_1S_2u = \nu u$ and $\nu > 0$. It is easy to show that $\lambda = \pm\nu^{1/2}/2$, $u_1 = u$, $u_2 = \pm S_2u/\sqrt{\nu}$ satisfy the system of equations for $T=0$. □

3.3.5

According to the theorem of Subsection 3.3.2, the characteristic function of the distribution of the inner product (x_1, x_2) has the form

$$\varphi(t) = \prod_{k=1}^{\infty} (1 - 2it\lambda_k)^{-1/2}, \tag{3.39}$$

where the λ_k are the eigenvalues of the system (3.38).

In the case of independence of x_1 and x_2, taking into account the lemma of Subsection 3.3.4, we obtain from (3.39)

$$\varphi(t) = \prod_{k=1}^{\infty} (1 + \nu_k t^2)^{-1/2}, \tag{3.40}$$

where the ν_k are eigenvalues of the operator S_1S_2.

In some particular cases the infinite product in (3.40) can be expressed in terms of elementary functions. This will be illustrated by the following example. Let us take, in any basis, diagonal matrices for covariance operators of independent Gaussian elements x_1 and x_2, and let the diagonal elements be equal to ak^{-2} and bk^{-2} ($k = 1, 2, \ldots$), respectively. Then, according to (3.40), the characteristic function of the inner product (x_1, x_2) has the form

$$\varphi(t) = \prod_{k=1}^{\infty} \left(1 + \frac{ab}{k^4} t^2\right)^{-1/2}, \tag{3.41}$$

and, by a simple transformation of a known formula, (3.41) becomes

$$\varphi(t) = \pi \left[\frac{\sqrt{4abt^2}}{\cosh\left(\pi^4\sqrt{4abt^2}\right) - \cos\left(\pi^4\sqrt{4abt^2}\right)} \right]^{1/2}.$$

3.3.6

Finally, we come to the problem of the integration of operator-valued functions with respect to measures whose values are random elements in H. Such integrals can be called *stochastic*. They arise naturally in the investigation of stochastic differential equations in Hilbert spaces.

The most important case is the one in which the functions are determined on an interval of the real line. But this restriction does not simplify the problem, and therefore we shall consider a more general case; we shall assume that the integrable functions are defined on some measurable space $(\Lambda, \mathfrak{B}(\Lambda))$. We have, as always, some basic probability space $(\Omega, \mathfrak{B}(\Omega), P)$ on which random elements are defined with values in H. Denote by $L_2(\Omega, \mathfrak{B}(\Omega), P; H)$ the Hilbert space of *random elements of second order in the strong sense*, that is, measurable mapping $x: \Omega \to H$ (there is a fixed σ-algebra of Borel sets in H), such that $E\|x\|^2 < +\infty$ (E means, as before, integration with respect to the measure P).

We shall deal with measures Φ, determined on the σ-algebra $\mathfrak{B}(\Lambda)$, taking values in the space $L_2(\Omega, \mathfrak{B}(\Omega), P; H)$ (countably additive in this space) and satisfying the following conditions.

a. $E(\Phi(\Delta), h) = 0$ for any $\Delta \in \mathfrak{B}(\Delta)$ and $h \in H$.

b. $E(\Phi(\Delta'), h)(\Phi(\Delta''), g) = 0$ for any h, $g \in H$ and disjoint Δ' and Δ'' in $\mathfrak{B}(\Lambda)$.

c. The linear operator S_Δ in H, determined by

$$(S_\Delta h, h) = E(\Phi(\Delta), h)^2,$$

has the form $S(\Delta) = \mu(\Delta)S$, where μ is a positive real function on $\mathfrak{B}(\Lambda)$ and S does not depend on Δ. It is easy to see that S is a nonnegative linear self-adjoint operator with a finite trace (S-operator) and that μ is an ordinary finite measure on $(\Lambda, \mathfrak{B}(\Lambda))$.

Now, we shall give the definition of a stochastic integral (with respect to measures that satisfy the preceding conditions)[11] of the functions $A(\lambda)$, $\lambda \in \Lambda$, whose values are linear operators that map the space H into itself. Further, we shall find the covariance operator of the stochastic integral as a random element in H, in the form of some strong integral (integral in the Bochner sense). Note that all of this can also be done for the case in which H is complex. The slight changes that have to be made are obvious.

---

[11] Measures that satisfy these conditions can be called *homogeneous orthogonal random measures of the second order with zero mean.*

3.3.7

We shall follow the general idea of defining a stochastic integral. First, assume that the function A is simple and denote by $A_1, A_2, \ldots, A_n$ the finite number of its values taken on measurable sets $\Delta_1, \Delta_2, \ldots, \Delta_n$. The stochastic integral J for a simple function A is determined in the usual way.

$$J = \int_\Lambda A(\lambda)\Phi\, d\lambda = \sum_{k=1}^{n} A_k \Phi(\Delta_k). \tag{3.42}$$

The problem is to extend this definition to a wider class of functions. Such an extension is possible by taking a suitable norm in the set of functions A. But first of all, one must choose a suitable norm in the set of values of the functions A. This norm should, of course, be connected with the measure Φ.

Let $N(A) = [Tr(ASA^*)]^{1/2}$, where A^* is the operator adjoint to A and S is the nonnegative operator with a finite trace that occurs in condition (c). Obviously, $N(A) < +\infty$ for each bounded linear operator A. Moreover, it is easy to show that the following inequality holds.

$$N(A) \leqslant (TrS)^{1/2} \|A\|, \tag{3.43}$$

where $\|A\|$ is the norm of the operator A in the usual sense. The function N also satisfies the following conditions.

$$N(A+B) \leqslant N(A) + N(B), \tag{3.44}$$

$$N(ASB) \leqslant \|S\|^{1/2} N(A) N(B). \tag{3.45}$$

Let $e_1, e_2, \ldots$ be any basis in H. After simple transformations, we obtain

$$N^2(A+B) = N^2(A) + N^2(B) + 2\sum_{k=1}^{\infty} (SB^* e_k, A^* e_k).$$

Now, using the inequality $(Sh, g)^2 \leqslant (Sh, h)(Sg, g)$, of which the proof is similar to the proof of the Cauchy-Bunyakovsky inequality,[12] we obtain

$$N^2(A+B) \leqslant N^2(A) + N^2(B) + 2\sum_{k=1}^{\infty} (SA^* e_k, A^* e_k)^{1/2} (SB^* e_k, B^* e_k)^{1/2},$$

which becomes condition (3.44) after applying Hölder's inequality. Now we shall prove condition (3.45). Applying once more the same inequalities, as

[12] In Western literature this inequality is called Schwarz's inequality.

well as inequality $(Sh,h) \leqslant \|S\|(h,h)$, we have

$$\begin{aligned}
N^2(AS) &= \sum_{k=1}^{\infty} (SSA^*e_k, SA^*e_k) \leqslant \|S\| \sum_{k=1}^{\infty} (SA^*e_k, SA^*e_k) \\
&\leqslant \|S\| \sum_{k=1}^{\infty} (SSA^*e_k, SA^*e_k)^{1/2}(SA^*e_k, A^*e_k)^{1/2} \\
&\leqslant \|S\| N(AS)N(A);
\end{aligned}$$

and we obtain the inequality $N(AS) \leqslant \|S\| N(A)$. Now, using this estimate, we obtain

$$\begin{aligned}
N^2(ASB) = Tr(ASBSB^*SA^*) &= \sum_{k=1}^{\infty} (BSB^*SA^*e_k, SA^*e_k) \\
&\leqslant \|BSB^*\| \sum_{k=1}^{\infty} (SSA^*e_k, A^*e_k) \\
&\leqslant \|BSB^*\| \sum_{k=1}^{\infty} (SSA^*e_k, SA^*e_k)^{1/2} \\
&\quad \times (SA^*e_k, A^*e_k)^{1/2} \leqslant \|BSB^*\| N(AS)N(A) \\
&\leqslant \|BSB^*\| \|S\| [N(A)]^2
\end{aligned}$$

and inequality (3.45) is established by noting that $\|BSB^*\| \leqslant Tr(BSB^*) = [N(B)]^2$.

Inequality (3.44), together with the relation $N(A) \geqslant 0$ and $N(\alpha A) = |\alpha| N(A)$ shows that N is a seminorm in the set of all bounded linear operators. But N is a norm only if none of the eigenvalues of the operator S are equal to zero. In the general case, after an obvious factorization, the function N will be a norm in the corresponding set of equivalence classes. This norm is suitable to define the stochastic integral. But when we compute the covariance operator of the stochastic integral, we would find difficulty in the fact that $N(A_k) \to 0$ does not imply $N(A_k^*) \to 0$. In order to avoid this difficulty, we take the function $n(A) = N(A) + N(A^*)$ instead of N. It is obvious that n is also a seminorm and the inequality (3.45) is also valid for n.

The set $\{A : n(A) = 0\}$ is a linear subspace of the linear space of all bounded operators A. The function n is a norm in the corresponding linear quotient space. Following tradition, we will not distinguish between the equivalence class and an individual operator from this class. In this sense we shall say that n is a norm in the linear space of all bounded linear operators. The completion of this space in the norm n will be denoted by $\mathscr{Q}_S$.

REMARK. The Banach space $\mathscr{Q}_S$ differs from the space of operators in the usual (uniform) metric; that is, it is not a ring, because the product AB is not determined for all classes A and B (for individual A and B from their classes, the product AB may belong in general to different classes). But because of inequality (3.45), the products of form ASB are well defined and belong to $\mathcal{Q}_S$ if $A, B \in \mathcal{Q}_S$.

3.3.8

Let us consider the Banach space $L_2 = L_2(\Lambda, \mathfrak{B}(\Lambda), \mu; \mathcal{Q}_S)$ of functions $A(\lambda)$, $\lambda \in \Lambda$ with values in $\mathcal{Q}_S$, measurable in the strong sense (in the Bochner sense), such that

$$|A|^2 = \int_\Lambda n^2(A(\lambda))\mu(d\lambda) < +\infty, \tag{3.46}$$

where μ is the measure in condition (c) in Subsection 3.3.6. The norm in L_2 is given by formula (3.46). (It is easy to see that this norm is generated by the inner product, so that L_2 is a Hilbert space, but this is not important to us here.)

It is known (see, e.g., reference 11, p. 125) that the set of all simple functions is everywhere dense in L_2. It is obvious that the set of simple functions whose values are bounded linear operators is also dense in L_2. For an arbitrary function of this type, the stochastic integral was determined in Subsection 3.3.7 as a finite sum (see (3.42)). This definition can be extended to all functions in the space L_2. To do this we need the following lemma.

Lemma. *The stochastic integral* (3.42) *is a random element in H with expectation zero and the covariance operator*

$$S_J = \sum_{k=1}^{n} A_k S A_k^* \mu(\Delta_k). \tag{3.47}$$

Further,

$$E\|J\|^2 \leqslant |A|. \tag{3.48}$$

PROOF. The fact that the expectation is zero follows from the definition and condition (a) for the measure Φ. Further, taking the other properties into account, we obtain

$$\begin{aligned}(S_J h, h) = E(J,h)^2 &= \sum_{i,k=1}^{n} E(A_i\Phi(\Delta_i),h)(A_k\Phi(\Delta_k),h)\\ &= \sum_{i,k=1}^{n} E(\Phi(\Delta_i),A_i^*h)(\Phi(\Delta_k),A_k^*h)\\ &= \sum_{i=1}^{n} E(\Phi(\Delta_i),A_i^*h)^2 = \sum_{i-1}^{n} (A_iSA_i^*h,h)\mu(\Delta_i).\end{aligned}$$

This formula is equivalent to (3.47), from which inequality (3.48) follows. This is true because the expectation of the square of the norm of the random element in H is equal to the trace of its covariance operator; and therefore,

$$E\|J\|^2 = \sum_{i=1}^{n} N^2(A_i)\mu(\Delta_i) \leqslant \sum_{i=1}^{n} n^2(A_i)\mu(\Delta_i) = |A|^2.$$

Now let A be an arbitrary function from the space L_2 and let $\{A_j\}$ be a sequence of simple functions converging to A. The sequence J_j of stochastic integrals of A_j by inequality (3.48) will be fundamental (Cauchy) in the metric of the space $L_2(\Omega, \mathfrak{B}(\Omega), P; H)$. Therefore it will be convergent to some random element J, which does not depend on the choice of the sequence $\{A_j\}$. This element will be called the *stochastic integral* over Λ of the function A with respect to the measure Φ.

Thus the stochastic integral is well defined and does exist for an arbitrary function $A \in L_2(\Lambda, \mathfrak{B}(\Lambda), \mu; \mathfrak{C}_S)$. □

3.3.9

Let us find the *covariance operator* S_J of the *stochastic integral* J. (It is obvious that the expectation of the S_J is zero.) The existence of S_J follows from the condition $E\|J\|^2 = |A|^2 < +\infty$. We shall show that

$$S_J = \int_\Lambda A(\lambda) S A^*(\lambda) \mu\, d\lambda, \tag{3.49}$$

where the integral is taken in the Bochner sense (the strong integral) relative to convergence in the space $\mathfrak{C}_S$. In order to prove (3.49) it is sufficient to establish the following: (1) the convergence of $\{S_{J_j}\}$ to S_J in the space $\mathfrak{C}_S$, and (2) the convergence in the measure μ of the sequence of simple functions $\{A_j(\lambda)SA_j^*(\lambda)\}$, $\lambda \in \Lambda$, to ASA^* (also in the norm of $\mathfrak{C}_S$). The first statement follows easily from $E\|J_j - J\|^2 \to 0$ and inequality (3.43). We shall prove statement (2). Using inequality (3.45), we obtain

$$\begin{aligned} n(A_jSA_j^* - ASA^*) &\leqslant n(A_jSA_j^* - A_jSA^*) + n(A_jSA^* - ASA^*) \\ &\leqslant \|S\|^{1/2}[n(A_j) + n(A)]n(A_j - A). \end{aligned}$$

It remains to note that convergence in measure follows from the convergence in L_2.

3.3.10

If $(\Lambda, \mathfrak{B}(\Lambda))$ is an interval on the real line with the Borel σ-algebra, then an important particular case of the previously described homogeneous orthogonal random measures of the second order with zero mean are measures having as "distribution functions" homogeneous Wiener processes with values in H. In this case the stochastic integral is a Gaussian random element in H with expectation zero, and its covariance operator completely determines its probability distribution.

Note that by a homogeneous Weiner process with values in H we can understand the family $W(t)$, $t_0 \leqslant t < t_1$, of random elements in H satisfying the following conditions:

a. $W(t_0) = 0$.
b. The increments on nonoverlapping intervals of the parameter are independent.
c. The increment $W(t+s) - W(t)$ has for all $t_0 \leqslant t < t+s < t_1$ a Gaussian distribution with expectation zero and covariance operator sS, where S does not depend on t or on s.

The existence of such a process can be shown by using Kolmogorov's theorem in its general form, as proved by I. I. Gihman and A. V. Skorohod [45].

4

Some Basic Problems in the Theory of Probability Distributions on Banach Spaces

Introduction

In this chapter we investigate problems connected with the basic notions of probability theory, such as the characteristic function, expectation, and variance as applied to random elements with values in arbitrary Banach spaces. Interpreted appropriately, these problems can be considered as problems of analysis in Banach spaces.

In addition to the usual difficulties connected with the fact that the space is infinite dimensional, difficulties also arise because the space is not self-adjoint (non-Hilbertian). This causes new problems that can be treated as problems of analysis in the broad sense and which may also be of some interest irrespective of their source.

The content of this chapter is more uniform, in comparison, for example, with the third chapter, and the material in each section can briefly be described.

In Section 4.1 we give the basic definitions and describe, in particular, the notion of a characteristic functional.

In Section 4.2 the Pettis integral is described. The integrability condition is given as a typical "condition in the weak sense," which is quite natural here. As a particular consequence of the fact that the space is not necessarily reflexive, it follows that the integrability over the whole space does not imply, in general, the integrability over all measurable subsets. A simple conclusion from the integrability condition is the existence of the expectation of an arbitrary Gaussian distribution on any separable Banach space.

In Section 4.3 the covariance operator is defined. It is shown that the condition for the existence of a covariance operator can be reduced to the natural necessary condition. Then, the basic properties of the covariance operators are given. The factorization lemma is proved here.

In Section 4.4 the problem of characterizing the class of all covariance operators of all probability distributions, as well as the subclass of Gaussian covariances, is considered.

The basic results of this chapter were published in references 46 to 48. The articles by Le Cam [49] and V. N. Sudakov and A. M. Vershik [50] should be mentioned in connection with some of the particular results of this chapter. U. Grenander [51] presents a survey of the results on distributions on Banach spaces, which were known in 1963.

4.1 Introductory Remarks: Fourier Analysis in Banach Spaces

4.1.1

Denote by $(\Omega, \mathfrak{B}(\Omega), P)$, as before, some basic probability space; that is, a space with a finite measure and with the additional property that the measure is normed, namely, $P(\Omega) = 1$. Further, let X be a real Banach space and $x: \Omega \to X$ be a mapping of Ω into X. We shall always consider weakly measurable mappings. The mapping x is called *measurable in the weak sense* if for all $f \in X^*$ the numerical functions $f(x(\cdot))$ are measurable (with respect to the class of Borel sets on the real line and the σ-algebra $\mathfrak{B}(\Omega)$ in Ω). This condition of measurability imposes a comparatively weak restriction on the mapping, and, moreover, is convenient in applications, because it allows us to reduce the problems to the respective one-dimensional variants.

The weakly measurable mapping $x: \Omega \to X$ is called in probability theory a *random element with values in* X. With each random element x there is associated a normed measure $\mu = \mu_x$ on the space X, the *distribution of the random element* x. Before giving a precise definition of this notion, we shall describe the σ-algebra on which the distribution μ is defined. This will be the σ-algebra induced by half-spaces of the form

$$\{x : f(x) < t\}, \qquad f \in X^*, \quad -\infty < t < +\infty, \tag{4.1}$$

that is, the smallest of all σ-algebras that contain sets of form (4.1) for all $f \in X^*$ and all real t. Let $\mathfrak{L}$ denote this σ-algebra. It is obvious that all linear continuous functionals in X are measurable with respect to $\mathfrak{L}$ (and to the σ-algebra of Borel sets of the real line) and that $\mathfrak{L}$ is the smallest of all the σ-algebras that have this property. From this it follows, in particular, that $\mathfrak{L} \subset \mathfrak{B}(X)$, where $\mathfrak{B}(X)$ denotes the σ-algebra of Borel sets (with respect to the metric topology) of the space X. It is well known that if X is separable, then the converse is also true, and hence in that case we have $\mathfrak{L} = \mathfrak{B}(X)$. This fact is rather important for problems of measurability and its proof is simple. We shall now give the proof.

Because of separability, each open set in X can be represented as a union of not more than a countable number of balls. Therefore it is enough to show that all the balls in the space X belong to $\mathfrak{L}$ (i.e., to show the measurability of all functionals $\|x - a\|$ for all fixed $a \in X$). Let $\{x_i\}$ be a countable set, everywhere dense in X and let f_i $(i = 1, 2, \ldots)$ be a linear functional, such that $f_i(x_i - a) = \|x_i - a\|$, $\|f_i\| = 1$ (the existence of f_i is one of the consequences of the Hahn–Banach theorem about the extension of the linear functionals). Together with the functionals f_i, the functional $\sup_i |f_i(x - a)|$ is also measurable, and therefore for a ball $\{x : \|x - a\| \leqslant r\}$ to belong to the σ-algebra $\mathfrak{L}$, we need only to prove that the following relation is true.

$$\{x : \|x - a\| \leqslant r\} = \left\{x : \sup_i |f_i(x - a)| \leqslant r\right\} \tag{4.2}$$

The fact that the left-hand side is a subset of the right-hand side is obvious. We shall show the opposite. If $\|x - a\| > r$, then too, $\|x_i - a\| > r$ for some i, because x_i can be taken close enought to x. But then, $f_i(x_i - a) = \|x_i - a\| > r$ and also $f_i(x - a) > r$. Thus if x belongs to the set on the right side of (4.2), then it also belongs to the left side. This ends the proof.

Therefore if the space X is separable, then the class of all half-spaces and the class of all balls give the same σ-algebra.

4.1.2

A random element can be defined as a measurable mapping of the basic probability space into the measurable space $(X, \mathfrak{L})$. It is obvious that this definition coincides with the previous definition of weak measurability. The distribution $\mu = \mu_x$ of the random element x is the usual (positive, countable additive) normed measure on the measurable space $(X, \mathfrak{L})$, given by

$$\mu_x(E) = P\{\omega : x(\omega) \in E\}, \qquad E \in \mathfrak{L}.$$

Conversely, each normed measure μ on $(X, \mathfrak{L})$ is the distribution of some random element (more accurately, of the family of identically distributed random elements), which, for example, can be taken as the identity mapping of the space X into itself. From the point of view of the problems ordinarily investigated in probability theory, a random element is completely characterized by its distribution. Therefore there exist two equivalent descriptions either in terms of random elements or in terms of distributions. There are no serious reasons for choosing one of these methods over the other; however, some choice is necessary, although no general choice needs to be made for all cases. Hence we shall pursue both possibilities.

Note the simple well-known formula by which we establish the relation between these two descriptions and which will be used later without additional explanation. Let F be an arbitrary measurable functional in X and x a random element with the distribution μ. The following equality for the two integrals holds.

$$\int_{\Omega} F(x(\omega))P(d\omega) = \int_{X} F(x)\mu(dx), \tag{4.3}$$

and either integral exists if and only if the other does.

Recall that E and E_{μ} means, respectively, integration with respect to the measure P over the whole set Ω and with respect to the measure μ over the whole X.

4.1.3

Although for general Banach spaces the two basic theorems (Bochner's and Levy's) that allow application of Fourier's method of transforms depart essentially from the classical finite-dimensional versions, the other basic properties hold, and therefore the Fourier analysis—the method of characteristic functionals—is still very important.

The *characteristic functional*, that is, the Fourier transform of the distribution μ (or of the random element with distribution μ) is a functional in the conjugate space X^*, defined by

$$\chi(f) = \chi(f;\ \mu) = E_{\mu} e^{if(x)}, \qquad f \in X^*.$$

Note the following basic properties of characteristic functionals, which hold in the general case (without any additional assumptions about the space X or the distribution μ).

1. The functional χ is positive definite and $\chi(0) = 1$.
2. The functional χ is uniformly continuous in norm in X^*.
3. If the sequence $\{\mu_n\}$ is weakly convergent to μ (i.e., $E_{\mu_n} F(x) \to E_{\mu} F(x)$ for each bounded continuous functional F on X), then $\chi(f;\ \mu_n) \to \chi(f;\ \mu)$.
4. $\chi(f;\ \mu_1 * \mu_2) = \chi(f;\ \mu_1)\chi(f;\ \mu_2)$. Here $\mu_1 * \mu_2$ denotes the ordinary convolution of the distribution μ_1 and μ_2, that is,

$$(\mu_1 * \mu_2)(E) = E_{\mu_1} \mu_2(E - x) = E_{\mu_2} \mu_1(E - x)$$

for each set $E \in \mathfrak{L}$ (supposing the existence of the convolution).

5. For each natural n and arbitrary $f_1, f_2, \ldots f_n$ from X^*, the function of n real variables

$$\varphi(t_1, t_2, \ldots, t_n) = \chi(t_1 f_1 + t_2 f_2 + \cdots t_n f_n)$$

is the characteristic functional of the distribution $\mu^{(f_1,f_2,\ldots,f_n)}$, which is the distribution of the random vector $((f_1,x),\ldots,(f_2,x),\ (f_n,x))$, determined on the basic probability space $(X,\mathfrak{L},\ \mu)$.

6. The distribution is uniquely determined by its characteristic functional; that is, if $\chi(f;\ \mu_1)=\chi(f;\ \mu_2)$ for all $f\in X^*$, then $\mu_1=\mu_2$ everywhere on $\mathfrak{L}$.

The proofs of properties (1) to (4) do not differ from the proofs of the corresponding statements in the finite-dimensional case. Property (5) follows easily from the definitions of χ and the characteristic function of an n-dimensional random vector. We shall now prove property (6). The σ-algebra $\mathfrak{L}$ is generated by half-spaces

$$\{x : f(x) < u\} \qquad f\in X^*, \quad -\infty < u < +\infty,$$

and therefore the distribution will be completely known if the measures of all these half-spaces are known. According to the one-dimensional variant of property (6), the measures of these half-spaces can be considered known, because the characteristic functions $\varphi_f(\cdot)$ of the random variables f are known. This follows because, from property (5), we have $\varphi_f(t)=\chi(tf)$, $f\in X^*$.

Note one more property that strengthens property (2).

7. The characteristic functional of an arbitrary distribution μ in X is sequentially continuous in the X-topology of the space X^*. The *X-topology* of the space X^* is determined by the base of neighborhoods of zero consisting of all sets of the form

$$\{f : |f(x_i)| < \epsilon, \quad i=1,2,\ldots,n\}, \quad \epsilon>0, \quad n\geqslant 1, \quad x_i\in X^*.$$

This topology is also called the *weakest* or the *weak * topology*. It is, of course, weaker than the metric topology (it is also weaker than the weak topology) of the space X^*.

The proof of property (7) follows easily from Lebesgue's theorem on convergence of integrals. Let $\{f_n\}$ be an arbitrary sequence of elements of the space X^*, convergent to the element f. One has to show that the sequence of complex numbers $\chi(f_n;\ \mu)$ is convergent to $\chi(f;\ \mu)$. For each value of the parameter n the complex function $\exp\{if_n(x)\}$ is measurable and bounded. Therefore it can be treated as an element of the Lebesgue space $L_1(X,\mathfrak{L},\ \mu)$. The sequence $\exp\{if_n(x)\}$ of the elements of this space is uniformly bounded and convergent everywhere in X to the function $\exp\{if(x)\}$; and one has only to apply the afore-mentioned Lebesgue theorem.

4.1.4

The characteristic functional can also be defined explicitly in terms of a random element without using the notion of distribution. If $x(\omega), \omega \in \Omega$ is a random element with values in X, then we define

$$\chi(f; x) = Ee^{if(x(\omega))}.$$

According to formula (4.3), if μ means the distribution of the random element x, then

$$\chi(f; x) = \chi(f; \mu). \tag{4.4}$$

The properties of the characteristic functionals from the previous subsection obviously also hold for the functionals $\chi(\cdot\,; x)$. Some properties need to be reformulated. We shall give a new form of property (4), but we first continue the numeration.

8. If x_1 and x_2 are independent random elements with values in the space X, then

$$\chi(f; x_1 + x_2) = \chi(f; x_1)\chi(f; x_2).$$

The fact that the sum of the two random elements with values in a separable Banach space is a random element can easily be established (recall that a random element is defined as a weakly measurable mapping).

The independence of random elements x_1 and x_2 is defined by the usual condition

$$P\{\omega : x_1(\omega) \in E_1, x_2(\omega) \in E_2\} = P\{\omega : x_1(\omega) \in E_1\}P\{\omega : x_2(\omega) \in E_2\}$$

for all $E_1, E_2 \in \mathfrak{L}$. It is sufficient that this condition be satisfied on all half-spaces of X, and therefore the definition of independence of the random variables x_1 and x_2 is equivalent to the independence of all random variables $f_1(x_1(\cdot))$ and $f_2(x_2(\cdot))$ for all $f_1, f_2 \in X^*$. In particular, if x_1 and x_2 are independent, then the random variables $f(x_1(\cdot))$ and $f(x_2(\cdot))$ are also independent, and property (8) is proved. From this it also follows that according to formula (4.4) and properties (4) and (6) for characteristic functionals, the distribution of the sum of independent random elements is equal to the convolution of the distributions of the summands.

Let us also note the formula for the transformation of the characteristic functional by a linear transformation of the random element. Let A be a bounded linear transformation of one Banach space X into another one, say Y. It is obvious that for each random element $x \in X$, the element $y(\omega) = Ax(\omega)$ is always a random element with values in Y. To prove this, one should show measurability, which can easily be done by using the relation $g(Ax) = (A^*g)(x)$, which determines the operator A^*, adjoint to A. Similarly, the following formula can be proved.

9. $\chi(g; Ax) = \chi(A^*g, x), \qquad g \in Y^*.$

4.1.5

Thus most of the basic properties of characteristic functions are preserved for the case of distributions on general Banach spaces. Two additional important theorems for the finite-dimensional case are the theorems of Bochner and Lévy. The first states that if the functional χ has properties (1) and (2), then it is the characteristic functional of some probability distribution. The second theorem states the converse to property (3).

In the first three chapters we spoke about some infinite-dimensional analogues to these important theorems for different spaces.

4.1.6

If the Banach space X is conjugate to some Banach space $Y, X = Y^*$, then, in addition to the previous definition, one often uses another, narrower definition of the characteristic functional of a distribution on X (or a random element with values in X). From the general definition, the characteristic functional χ is determined in the space $X^* = Y^{**}$ and the restriction of χ to $Y \subset Y^{**}$ is the characteristic functional in the narrower definition.

As already known, an element $y \in Y$ can be regarded as a linear functional $y(g) \equiv g(y)$ in Y^*, and this natural inclusion of Y into Y^{**} is an isometric isomorphism. Therefore it is not necessary to distinguish between the space Y and its image in Y^{**}. The inclusion $Y \subset Y^{**}$ will be understood (here and elsewhere) in this sense.

Obviously, it is not necessary first to define the characteristic functional in the wide sense and then to take its restriction. From the beginning it is possible to introduce in the space Y^* a σ-algebra $\mathfrak{L}_0$ generated by half-spaces $\{g : y(g) < t\}$, which are determined by linear functionals belonging to the space Y (more accurately, to the image of Y in Y^{**} by the natural inclusion). It is obvious that $\mathfrak{L}_0 \subset \mathfrak{L}$ and if the space is reflexive, then $\mathfrak{L}_0 = \mathfrak{L}$, where $\mathfrak{L}$ again means the σ-algebra generated by all half-spaces (the equality $\mathfrak{L}_0 = \mathfrak{L}$ is also possible in the nonreflexive case; as an example one can take the space ℓ_1).

Treating measurability (particularly weak measurability of the mapping of Ω into Y^*) as measurability with respect to the σ-algebra $\mathfrak{L}_0$, we can give the new definition of the characteristic functional of the distribution μ in

Y^* by

$$\chi_0(y;\mu) = E_\mu e^{ig(y)}, \qquad y \in Y.$$

Properties (1) to (7) still hold for the characteristic functionals χ_0 after some obvious modifications. Moreover, defining the characteristic functional $\chi_0(\cdot; g)$ of the random element g as a weakly measurable (with respect to the σ-algebra $\mathfrak{L}_0$) mapping $\Omega \to Y^*$, we see that the analogues of properties (8) and (9) also hold for χ_0. Thus the functional χ_0 is more convenient and as applicable as the functional χ. But not every space is a conjugate space; and therefore, in general, one has to use the characteristic functional in the wide sense, although in some particular cases we can use the simpler characteristic functionals in the narrow sense. This was done in the first chapter, taking for the spaces ℓ_1 and ℓ_∞ the σ-algebra $\mathfrak{L}_0$ and the characteristic functional χ_0. But for the space c_0 we took the σ-algebra $\mathfrak{L}$ and characteristic functional χ.

If the space is reflexive, then these two methods are identical and $\chi_0 = \chi$. In the general case χ_0 is a restriction of χ, that is, $\chi_0(u;\mu) = \chi(u;\mu)$ for $u \in Y$. The following theorem shows that in the separable case the functional χ is uniquely determined by its restriction χ_0.

Theorem. *Let Y^* be separable and μ_1 and μ_2 be distributions on a measurable space $(Y^*, \mathfrak{L})$. If $\chi(u;\mu_1) = \chi(u;\mu_2)$ for all $u \in Y$, then $\chi(u;\mu_1) = \chi(u;\mu_2)$ everywhere in Y^{**} (and therefore $\mu_1 = \mu_2$ everywhere on $\mathfrak{L}$).*

PROOF. Denote by S and S^{**} balls with centers at the origin and an arbitrarily given radius r in the spaces Y and Y^{**}, respectively. It is known ([11], p. 424), that S is everywhere dense in S^{**} in the Y^*-topology of the space Y^{**}. On the other hand, because of the separability of Y^*, this topology on S^{**} is metrizable ([11], p. 426). Therefore according to property (7) of characteristic functionals, the equality $\chi(f;\mu_1) = \chi(f;\mu_2)$ holds for all $f \in S^{**}$ (and hence everywhere in Y^{**}) because radius r has been chosen arbitrarily. □

4.1.7

Finally, we shall formulate the following basic supposition, which is fundamental for all further investigation. Assume that the distributions considered and the corresponding random elements are of the second order in the weak sense, that is, they satisfy the condition for all $f \in X^*$:

$$E_\mu f^2(x) < +\infty$$

and, correspondingly,

$$Ef^2(x(\omega)) < +\infty. \tag{4.5}$$

This restriction is much weaker than the condition of second order in the strong sense.

$$E_\mu \|x\|^2 < +\infty$$

and, correspondingly,

$$E\|x(\cdot)\| < +\infty. \tag{4.6}$$

Condition (4.6) obviously implies condition (4.5). The converse is not true. Moreover, condition (4.5) does not imply condition (4.6), even in the case of a Hilbert (infinite-dimensional) space, even if the weak integrability occurs for any positive power and the strong integrability occurs in the first power only. This is shown by the following example.

EXAMPLE. Let $X = \ell_2$, $\Omega = \{\omega_1, \omega_2, \ldots\}$, and $\mathfrak{B}(\Omega)$ be the σ-algebra of all subsets. The probability P and the random element x is defined by $P(\omega_k) = P_k$, $P_k = Ck^{-1}(\ln k)^{-2}$, $(k = 1, 2, \ldots)$, where C is the norming constant and $x(\omega_k) = x_k$, where $x_j^k = 0$ for $k \neq j$ and $x_{k,k} = \ln k$. One can easily show that for each $h \in l_2$ the inner product (h, x) is integrable with respect to P with an arbitrary positive power, but $\|x\|$ is not integrable.

4.2 The Pettis Integral and the Expectation of Random Elements

4.2.1

The expectation of a random element $x: \Omega \to X$ obviously should be defined as the integral with respect to the measure P of the function x. But the integral of vector-valued functions can be understood in different ways. It is known that for this purpose the most convenient kind of integral is the *weak* or *Pettis integral*, which we shall now define and investigate.

The *Pettis integral* of a weakly measurable mapping $x: \Omega \to X$ with respect to the measure P is an element $m \in X$ (if it exists), which satisfies the condition

$$f(m) = \int_\Omega f(x(\omega)) P(d\omega) \qquad \text{for all } f \in X^*.$$

Remembering the probabilistic interpretation of this integral, we denote the element m by $Ex(\omega)$ or, more simply, by Ex.

Note the basic properties of the expectation (or the Pettis integral).

1. If Ex exists, then it is uniquely determined.

This follows from the fact that if $f(m) = 0$ for all $f \in X^*$, then $m = 0$ according to one of the corollaries of the Hahn–Banach theorem.

2. If Ex_1 and Ex_2 exist, then for all constants c_1 and c_2 the linear

combination $c_1x_1 + c_2x_2$ is integrable and

$$E(c_1x_1 + c_2x_2) = c_1Ex_1 + c_2Ex_2.$$

3. If $P\{\omega : x(\omega) = c\} = 1$, then $Ex = c$ (c is constant).
4. If $X = R^1$, then the Pettis integral coincides with the Lebesgue integral.

These statements follow immediately from the definition of the Pettis integral.

5. Let A be a bounded linear mapping of the space X into some other space Y, and let Ex exist. Then $E(Ax)$ also exists and is equal to $A(Ex)$.
6. If $\|x\|$ is measurable and $E\|x\| < +\infty$, then Ex exists and the following estimate holds.

$$\|Ex\| \leqslant E\|x\|.$$

Property (5) can easily be proved by using the adjoint operator and the formula defining it. The proofs for properties (5) and (6) are given by Hille and Phillips [52]. E. Mourier [3] has presented a probabilistic proof (based on the law of large numbers for random elements in X) of property (6), but with the assumption that X is reflexive and separable. Note also that the proof of this property for the separable case is given in Subsection 4.2.3 (see footnote 1).

The notion of expectation is a natural generalization of the notion of expectation of a finite-dimensional random vector $x = (x_1, x_2, \ldots, x_k)$. In this case Ex is defined as the vector $Ex = (Ex_1, Ex_2, \ldots, Ex_k)$, if all Ex_k exist. In the infinite-dimensional case, if we assume that there exists a basis and the element x can be written as a sequence of its coordinates, then, in addition to the existence of Ex_j for all j, one has to require that the element with coordinates Ex_j belongs to the space X.

4.2.2

The integrability of all bounded linear functionals, that is, the condition

$$f(x(\cdot)) \in L_1(\Omega, \mathfrak{B}(\Omega), P) \qquad \text{for all } f \in X^* \tag{4.7}$$

is the natural necessary condition for the existence of the Pettis integral, because without this condition the definition of the integral does not make sense.

I. M. Gelfand and N. Dunford have shown (independently) that under assumption (1), used in the definition of Ex, the integral

$$\int_{\Omega} f(x(\omega))P(d\omega),$$

considered as a function of $f \in X^*$ is continuous with respect to the norm of the space X^*. Therefore it can be treated as an element of the second conjugate space X^{**}, because the additivity is obvious (the proof can be found in reference 52, p. 77). From this it follows that if the space X is reflexive, then the necessary condition (4.7) is also sufficient. This can be written as a theorem.

Theorem. *If the space X is reflexive, then condition* (4.7) *is necessary and sufficient for the existence of the Pettis integral. In particular, each random element that is weakly second order in a reflexive Banach space possesses an expectation.*

4.2.3

The concept of expectation is one of the very basic notions in the theory of probability, and it would be nice to have the condition of its existence without the assumption of reflexivity. This is needed to include the case of the space of continuous functions, one of the most important in analysis, particularly in the theory of probability. The importance of this space comes not only from the fact that it is the natural sample space for stochastic processes with continuous paths but also from the well-known fact (*Banach–Mazur theorem*) that an arbitrary separable Banach space can be isometrically and isomorphically embedded into the space of continuous functions.

According to the condition of strong continuity in f of the functional $Ef(x(\omega))$, given in section 4.1, the problem of finding conditions for the existence of the Pettis integral can be reduced to the problem of finding the conditions under which the integral $Ef(x(\omega))$, treated as an element of the space X^{**}, belongs to the image of the space X by the natural embedding of X into X^{**}. Therefore we use Banach's theorem, which states that an element $u \in X^{**}$ belongs to the image of X in the natural embedding of X into X^{**}, if the linear bounded functional $u(f)$, $f \in X^*$, is continuous in the X-topology of the space X^* (the definition of this topology has been referred to in Subsection 4.1.3). For the case when X is separable, this condition means that from the convergence $f_n(x) \to f(x)$ for all $x \in X$ the convergence $u(f_n) \to u(f)$ follows (Banach's theorem for this case is given in Reference 53, p. 131). Thus Banach's theorem essentially reduces the problem to one of finding the condition that permits passage to the limit under the integral $Ef(x(\omega))$, having pointwise convergence of the sequence of integrands. But the known conditions for passage to the limit under the integral sign do not lead to new interesting formulations.[1] Condition (4.8) is imposed for this purpose and gives the condition of integrability in the Pettis sense in natural terms. Assume that the space X is separable.

Theorem. *The Pettis integral with respect to the measure P of a weakly measurable mapping $x:\Omega\to X$ exists, if in addition to the natural necessary condition* (4.7), *the following condition also holds*: *There exists a number* $\epsilon>0$, *such that*

$$\inf_{f\in X^*} P\left\{\omega: |f(x(\omega))| \geqslant \epsilon\left|\int_\Omega f(x(\omega))P(d\omega)\right|\right\} > 0. \tag{4.8}$$

PROOF. Denote for brevity the integral $Ef(x(\omega))$ by $I(f)$. According to previous considerations, it is sufficient to prove that $I(f_n)\to 0$ for every sequence $f_n \in X^*$ $(n=1,2,\dots)$ of linear functionals on X that are weakly convergent to zero (because it is sufficient to check continuity at zero). Assume the converse, then there exists a number $\delta>0$ and a subsequence $\{f_{n_j}\}$, such that

$$|I(f_{n_j})| \geqslant \delta, \qquad j=1,2,\dots. \tag{4.9}$$

Now, let

$$A_j = \left\{\omega: |f_{n_j}(\omega)| \geqslant \epsilon |I(f_{n_j})|\right\} \tag{4.10}$$

and put

$$A = \limsup_{j\to\infty} A_j = \bigcap_{n=1}^{\infty}\bigcup_{j=n}^{\infty} A_j,$$

the limit superior of the sequence of the sets A_j. We shall obtain the needed contradiction if we can show that set A is not empty. For if $\omega_0\in A$ and $x_0 = x(\omega_0)$, then according to the definition of the upper limit of the sequence of sets, and according to (4.9) and (4.10), we would have the inequality

$$|f_{n_j}(x_0)| \geqslant \epsilon|(f_{n_j})| \geqslant \epsilon\delta$$

---

[1] Note that in the separable case the application of Lebesgue's theorem on putting the limit under the integral sign ([11], p. 124) gives at once a simple proof of the existence of Ex from the condition $E\|x\| < +\infty$; and hence gives the proof of property (6) of expectations, because the estimate $\|Ex\| \leqslant E\|x\|$ follows immediately from Hölder's inequality $|f(x(\omega))| \leqslant \|f\|\,\|x(\omega)\|$.

for an infinite set of numbers j. But this contradicts the weak convergence to zero of the initial sequence of linear functionals $\{f_n\}$. Thus we must establish that set A is nonempty. It follows from condition (4.8) that we have $P(A) > 0$, because for an arbitrary (finite) measure P the following inequality holds.

$$P\left(\limsup_{j\to\infty} A_j\right) \geqslant \limsup_{j\to\infty} P(A_j). \qquad \square$$

4.2.4

After the last conjecture one generally expects that condition (4.8), whose sufficiency has just been shown, is not necessary for the existence of the Pettis integral. This will be shown by the following example. But the example in Subsection 4.2.5 shows that this condition is usual for integrability in the weak sense. Strong integrability (in the Bochner sense), which is equivalent to the integrability of the norm $\|x\|$ does not follow from this condition.

EXAMPLE. Let $\Omega = \{\omega_n : n = 1, 2, \ldots\}$, let $\mathfrak{B}(\Omega)$ be the σ-algebra of all subsets of Ω and let the measure P be defined by

$$P(\omega_n) = p_n = 6/(n^2\pi^2), \qquad (n = 1, 2, \ldots).$$

Let us take the Hilbert space ℓ_2 as X and define the function $x : \Omega \to \ell_2$ in the following way.

$$x(\omega_n) = x_n, \qquad x_{n,k} = 0 \quad \text{if } k \neq n \quad \text{and} \quad x_{n,n} = n.$$

Further, let e_n be a linear functional in l_2, which is determined by the coordinates $e_{n,k} = 0$ for $k \neq n$ and $e_{n,n} = 1$, $(n = 1, 2, \ldots)$. It is obvious that for each n the function $e_n(\omega) = e_n(x(\omega))$ takes two values: $e_n(\omega_k) = 0$ for $k \neq n$ and $e_n(\omega_n) = n$. Therefore

$$\int_\Omega e_n(\omega) P(d\omega) = np_n,$$

and for any $\epsilon > 0$

$$P\{\omega : |e_n(\omega)| \geqslant \epsilon np_n\} = P\{\omega : e_n(\omega) = n\} = p_n \to 0$$

for $n \to \infty$. Therefore condition (4.8) is not satisfied. But it is easy to show explicitly that the Pettis integral exists in this case (and is equal to the element $\{m_k\}$, where $m_k = 6/\pi^2 k$, $k = 1, 2, \ldots$).

4.2.5

EXAMPLE. Let Ω and $\mathfrak{B}(\Omega)$ be determined as in Example 4.2.4. Take as X the Banach space c_0 (of real numerical sequences $x = \{x_k\}$ convergent to zero, with the norm $\|x\| = \sup_k |x_k|$). Define the function $x : \Omega \to X$ in the following way: $x(\omega_n) = x_n$, $x_n = \{x_{n,k}\}$, where for $n = 2j - 1$ we have $x_{2j-1,k} = 0$ for $k \neq j$ and $x_{2j-1,k} = j^2$; whereas for $n = 2j$ we have $x_{2j,k} = 0$ for $j \neq k$, $x_{2j,j} = -j^2$, $(j = 1, 2, \ldots)$. Define the measure P by

$$P(\omega_n) = p_n, \qquad p_{2j-1} = p_{2j} = 6/(\pi^2 j^2), \quad (j = 1, 2, \ldots).$$

Each linear functional in c_0 has the form

$$f(x) = \sum_{k=1}^{\infty} f_k x_k,$$

where

$$\sum_{k=1}^{\infty} |f_k| < +\infty.$$

Therefore we have

$$\begin{aligned}
\int_\Omega f(x(\omega)) P(d\omega) &= \sum_{j=1}^{\infty} f(x(\omega_{2j-1})) p_{2j-1} + \sum_{j=1}^{\infty} f(x(\omega_{2j})) p_{2j} \\
&= (6/\pi^2) \sum_{j=1}^{\infty} f(x_{2j-1}) j^{-2} + (6/\pi^2) \sum_{j=1}^{\infty} f(x_{2j}) j^{-2} \\
&= (6/\pi^2)\left(\sum_{j=1}^{\infty} f_j - \sum_{j=1}^{\infty} f_j \right) = 0,
\end{aligned}$$

and the necessary condition (4.7) and condition (4.8) are obviously satisfied. But the Bochner integral does not exist, because integrability in the Bochner sense is equivalent to the integrability of the norm and

$$\begin{aligned}
\int_\Omega \|x(\omega)\| P(d\omega) &= \sum_{j=1}^{\infty} \|x(\omega_{2j-1})\| p_{2j-1} + \sum_{j=1}^{\infty} \|x(\omega_{2j})\| p_{2j} \\
&= (6/\pi^2)\left(\sum_{j=1}^{\infty} 1 + \sum_{j=1}^{\infty} 1 \right) = +\infty.
\end{aligned}$$

Using these ideas, one can easily construct a similar example in the space ℓ_2. Thus integrability in the Bochner sense does not follow from conditions (4.8), (4.9), and (4.10) (not even in the case of a separable Hilbert space).

4.2.6

Now let x be a Gaussian random element with values in X. The probability that a Gaussian random variable is larger than its expectation is always $\frac{1}{2}$, independent of the variance. Therefore we have for all $f \in X^*$

$$P\{\omega : |f(x(\omega))| \geqslant |Ef(x(\omega))|\} \geqslant P\{\omega : f(x(\omega)) \geqslant Ef(x(\omega))\} = 1/2,$$

and condition (4.8) holds (take $\epsilon = 1$). The fact that the necessary condition (4.7) holds is obvious. Therefore the next theorem simply follows from the theorem of Subsection 4.2.3.

Theorem. *The expectation of an arbitrary Gaussian element with values in an arbitrary separable Banach space does exist.*

4.2.7

At the end of this subsection we shall make a remark that although not interesting from the point of view of probabilistic reasoning, is of some interest for the general theory of integration of vector-valued functions.

The definition of the Pettis integral has been given only for integration over the whole set Ω. Similarly, the Pettis integral can be defined over any measurable set.[2] We want to note that unlike the other definitions of an integral, and also unlike integration in the Pettis sense in the case of a reflexive space X, in the nonreflexive case integrability in the Pettis sense over Ω does not imply integrability over an arbitrary measurable subset $\Omega_1 \subset \Omega$. This can be demonstrated by Example from Subsection 4.2.5. Let $\Omega_1 = \{\omega_{2j-1}, j = 1, 2, \ldots\}$ and $f = \{f_k\} \in l_1$ be an arbitrary bounded functional in c_0. Using the notation from the Example 4.2.5, we have

$$\begin{aligned}\int_{\Omega_1} f(x(\omega))P(d\omega) &= \sum_{j=1}^{\infty} f(x(\omega_{2j-1}))p_{2j-1}\\ &= (6/\pi^2)\sum_{j=1}^{\infty} f(x^{2j-1})j^{-2}\\ &= (6/\pi^2)\sum_{j=1}^{\infty} f_j.\end{aligned}$$

This formula means that the integral

$$\int_{\Omega_1} f(x(\omega))P(d\omega),$$

[2] It is sufficient to take as the probability space the triple $(\Omega_1, \mathfrak{B}_1, P_1)$, where $\mathfrak{B}_1$ and P_1 are the obvious restrictions to Ω_1 of the σ-algebra $\mathfrak{B}(\Omega)$ and the measure P, respectively. It is also obvious that all the properties of Pettis integral are preserved in this case.

treated as a linear functional in $c_0^* = \ell_1$ is determined by the element $(6/\pi^2)(1, 1, \ldots, 1, \ldots)$ from c_0^{**}, which does not belong to c_0. Therefore the Pettis integral over the set Ω_1 does not exist, whereas the integral over Ω does exist (and is equal to the zero element of the space X).

Let us note that the theorem of Subsection 4.2.3 remains true for integration over an arbitrary set Ω_1 if one makes the obvious changes in condition (4.8). According to the previous statement, the fact that this condition is satisfied for the whole set Ω does not imply that it is satisfied for the subset Ω_1.

4.3 Operators Mapping Spaces into Their Conjugates and the Covariance of Random Elements

4.3.1

In the second chapter, the covariance matrix of the distributions (or random elements) was used quite often in different spaces of numerical sequences. They were mainly covariance matrices of Gaussian distributions and were determined in the natural way as simply infinite-dimensional analogues of covariance matrices of finite-dimensional random variables. By such a definition, the covariance matrices do not have to satisfy any properties, and therefore the problem of their existence is trivially solved. The covariance matrix of a random numerical sequence $\{x_k\}$ exists if and only if the condition $Ex_k^2 < +\infty$ is satisfied for all k (then, $E|x_k x_j| < +\infty$, too, according to the Cauchy–Bunyakovsky inequality).

The covariance operator in a Hilbert space cannot be defined as a simple analogue of the finite-dimensional covariance matrix. Now, we have to demand that the covariance matrix be the matrix of a linear (bounded) operator (because such a definition of covariance in Hilbert space is much more convenient). Therefore the covariance matrix of a distribution on the space ℓ_2, as in the space of sequences in general, does not give the covariance operator of a distribution on ℓ_2, if ℓ_2 is considered as a Hilbert space (the simplest example would be the distribution of a random element concentrated on ℓ_2 and consisting of pairwise noncorrelated random variables with variances finite but increasing to $+\infty$). Accordingly, the question of the existence of the covariance operator of distributions on Hilbert spaces H is nontrivial.

We shall show later, by using the closed graph theorem, that the natural necessary condition (second order in the weak sense) for the existence of the covariance operator as a bounded linear mapping $H \to H$ is also a sufficient condition. By the assumption of second order in the strong sense, the notion of a covariance operator of a random element in H has been given by E. Mourier and subsequently used by many authors. Only covariance operators with finite trace have been considered ([51], p. 128), because second order in the strong sense implies that the trace is finite. Actually, however, the class of covariance operators is much broader, and we shall make use of the fact that even the identity operator is a covariance operator.

These considerations hold not only for distributions on Hilbert spaces but for the more general idea of a covariance operator of probability distributions (or random elements) on arbitrary Banach spaces. In this general case we also assume only second order in a weak sense. By generalizing the case of a self-adjoint (Hilbert) space, we define the covariance operator as a bounded linear mapping of the conjugate space X^* into the second conjugate X^{**}. Before giving the exact definition, we would like to consider the properties of operators that map spaces into their conjugates. Such operators, unlike those that map in general one Banach space into another, have some of the properties that make them more like operators that map H into H. In particular, one can speak about such operators as being symmetric and nonnegative, and this is an important fact.

4.3.2

Keeping future applications in mind, we shall consider the case in which the space being mapped is itself a conjugate; that is, it is a mapping from X^* into X^{**}. Let R denote such a mapping. Then, for all $f, g \in X^*$ the expression $(Rf)(g)$ is well defined as the value of the linear functional $Rf \in X^{**}$ on the element $g \in X^*$, and the symmetry and nonnegativity of the operator R can be given by the relations

$$(Rf)(g) = (Rg)(f) \qquad \text{for all } f, g \in X^*,$$

$$(Rf)(f) \geqslant 0 \qquad \text{for all } f \in X^*.$$

The family of all symmetric and nonnegative mappings from X^* into X^{**} will be denoted by $\mathfrak{R}(X)$.

We shall now prove an important lemma on factorization, which may have interest beyond probability distributions on linear spaces.

Lemma. *Each operator $R \in \mathfrak{R}(X)$ can be expressed as $R = A^*A$, where A is a bounded linear mapping of the space X^* into an everywhere dense set of some Hilbert space H. This expression is unique up to a unitary equivalence, that is, if also $R = A_1^*A_1$, where A_1 maps X^* into an everywhere dense set of a Hilbert space H_1, then there exists a unitary operator $U: H_1 \to H$ and $A = UA_1$.*

PROOF.[3] It is obvious that the set $M = \{f:(Rf)(f) = 0\}$ is a subspace of X^* (it is also closed, but this is not important here). Let us consider the quotient space X^*/M with the usual definition of linear operations. Let $\tilde{f}$ denote the equivalence class that contains f. It is easy to show, by using the same technique as in the case when X is a Hilbert space, that the expression $(\tilde{f}, \tilde{g})$ given by the formula

$$(\tilde{f}, \tilde{g}) = (Rf)(g) \tag{4.11}$$

is well defined (i.e., the right-hand side of this formula depends only on the equivalence classes and does not change when the representatives change and defines the inner product in the linear space X^*/M, making it a pre-Hilbert space. The completion of this pre-Hilbert space is a Hilbert space, denoted by H. Further, let $A: f \to \tilde{f}$ be the natural homomorphism of X^* on X^*/M, considered as a mapping into H. It is obvious that A is linear and bounded and that AX^* is dense in H. Further, using (4.11) and the general definition of an adjoint operator, we obtain

$$(Rf)(g) = (Af, Ag) = (A^*Af)(g), \qquad f, g \in X^*,$$

from which the factorization $R = A^*A$ follows, as required.

We shall now show the uniqueness of the factorization. Suppose that in addition to $R = A^*A$ we also have $R = A_1^*A_1$, where A_1 maps X^* into a dense subset of a Hilbert space H_1. First, define the operator $U: H_1 \to H$ on the elements of this dense subset, taking $U(A_1 f) = Af$. The correctness of this definition follows from the equations

$$(A_1 f, A_1 f)_{H_1} = (Af, Af)_H = (Rf)(f), \quad f \in X^*, \tag{4.12}$$

from which $Af = 0$ if $A_1 f = 0$, and therefore if $A_1 f = A_1 g$, then $Af = Ag$. We note by (4.12) that the operator U is an isometry and therefore can be extended to a unitary operator $H_1 \to H$. Retaining U as the notation for this operator, we obviously have $UA_1 = A$. □

REMARK. If the space X is separable and if $RX^* \subset X$, then the Hilbert space H is separable.

In fact, in this case the operator $A: X^* \to H$ is continuous in the weak topology in H and in the X-topology in X^*. This can easily be shown as follows: If the generalized sequence $\{f_\alpha\}$ converges to zero in the X-topology, then for each $h \in H$ we have $(Af_\alpha, h) = f_\alpha(A^*h) \to 0$, because

[3] Our first proof was based on explicit considerations and required separability of the space X. It was S. A. Chobanyan who noted the possibility of using general considerations, and it is his proof that is presented here.

$f_\alpha(x) \to 0$ for all $x \in X$ and $A^*H \subset X$. Further, each closed ball S in X^* is compact in the X-topology ([11], p. 424), and if X is separable, then this topology on S is metrizable ([11], p. 426). From this it follows that in this case, X^* is separable in the X-topology; and from the continuity of the operator A, it follows that the space H is separable in the weak topology. Using the Hahn-Banach theorem, one can easily show that the separability of H in the weak topology is equivalent to the separability in the strong topology (this holds also for an arbitrary Banach space).

4.3.3

Now let us define the covariance operator. For this purpose we shall use the random element interpretation. Remember that we assume second order in the weak sense; that is, the following condition is satisfied.

$$Ef^2(x(\omega)) < +\infty \qquad \text{for every } f \in X^*. \tag{4.13}$$

Let us also assume that Ex exists (if X is reflexive, then the existence of Ex follows from condition (4.13)), and for brevity we take $Ex = 0$.

Consider the expression $R(f, g)$ given by the formula

$$R(f, g) = Ef(x(\omega))\,g(x(\omega)). \tag{4.14}$$

Because of condition (4.13) the functional $R(f, g)$ is determined everywhere in the Cartesian product $X^* \times X^*$. Further, it is obvious that $R(f, g)$ is a bilinear functional and that it is symmetric and nonnegative; that is,

$$R(f, g) = R(g, f), \qquad R(f, f) \geqslant 0, \quad f, g \in X^*. \tag{4.15}$$

We shall show that the bilinear functional $R(f, g)$ is bounded. To do this we use the closed graph theorem (see, e.g., reference [11], p. 57), according to which an everywhere defined and closed linear mapping of a Banach space into another one is a bounded mapping. Apply this theorem to the mapping $A : X^* \to L_2(\Omega, \mathfrak{B}(\Omega), P)$, which is determined by the formula $Af = f(x(\cdot))$. Because of condition (4.13), A is really a mapping of the space X^* into $L_2(\Omega, \mathfrak{B}(\Omega), P)$ determined everywhere in X. The linearity of the operator A is obvious. The fact that it is closed (also in a stronger sense than it is necessary) is a simple conclusion of the fact that if a sequence of functions from L_2 is pointwise convergent and also convergent in the L_2 metric, then the limiting functions are identical almost everywhere. Thus from the closed graph theorem the operator A is bounded; that is, there exists a constant $C > 0$, such that

$$Ef^2(x(\omega)) = \|Af\|^2 \leqslant C\|f\|^2. \tag{4.16}$$

If the right-hand side of (4.19) is estimated by Cauchy–Bunyakovsky's inequality and then estimate (4.16) is applied, we obtain the inequality $|R(f, g)| \leqslant C\|f\|\,\|g\|$, which means that the bilinear functional under consideration is bounded.

Now, similarly, as in the case of a Hilbert space, to each bounded bilinear functional there can be assigned a corresponding bounded linear operator. Fix the element g in $R(f, g)$ and consider R as a function of one variable in X^*. Then, we obtain a bounded linear functional in X^*, that is, an element in X^{**}. This element will depend on the fixed element $g \in X^*$, which means that we have defined a mapping of the space X^* into X^{**}. One can easily see that this mapping is defined everywhere, that it is linear and bounded. We will use the same letter R for this mapping as for the functional that generates it. Now, using relation (4.14), we can write the formula that determines the operator R as

$$(Rf)(g) = Ef(x(\omega))\, g(x(\omega)). \tag{4.17}$$

This operator $R = R_x$ will be called the *covariance operator* of the random element x. Note that if we do not assume $Ex = 0$, then instead of (4.17) we have the equality

$$(Rf)(g) = Ef(x(\omega) - Ex)\, g(x(\omega) - Ex). \tag{4.18}$$

4.3.4

From condition (4.15) or, explicitly, from the above formula, symmetry and nonnegativity of the covariance operators follow. We shall now give other properties.

Let x_1 and x_2 be independent random elements with value in X and let A be a bounded linear mapping of a Banach space X into a Banach space Y. The following formulas hold and their proofs follow easily from the definition of the covariance operator and the general definition of a conjugate operator.

$$R_{x_1+x_2} = R_{x_1} + R_{x_2},$$
$$R_{Ax} = A^{**} R_x A^*,$$

where A^{**} denotes the conjugate to A^*.

Note also that according to the obvious identity

$$4R(f, g) = R(f+g, f+g) - R(f-g, f-g),$$

which holds for each symmetric bilinear functional, the covariance operator is well determined by the corresponding quadratic functional. Therefore we may substitute the following expression for (4.18).

$$(Rf)(f) = Ef^2(x(\omega) - Ex).$$

This shows, in particular, that in the case of Hilbert spaces the definition of the covariance operator given here coincides with the one given in Subsection 3.1.4.

Concerning the connection with the definition of the covariance matrix for the case of Banach spaces of numerical sequences, we easily see that if the covariance operator exists, then its representation in the natural basis gives the covariance matrix. The existence of the covariance matrix does not imply the existence of the covariance operator (this is the case only with the additional necessary and sufficient condition of second order in the weak sense).

4.3.5

The nonnegative quadratic functional is called *nondegenerate*, if it is zero at the zero point only. Accordingly, the covariance operator R will be called *nondegenerate* if $(Rf)(f) = 0$ only for $f = 0$. On the other hand, it is natural to call the random element $x \in X$ *nondegenerate*, if there does not exist a hyperplane in X that contains all the values of the random element. This condition means that all random variables $f(x(\cdot))$, $f \in X^*$, are nondegenerate, that is, they are not concentrated at a point and hence have nonzero variances. But the variance of the random variable $f(x(\cdot))$ is $(Rf)(f)$, so that the nondegeneracy of the random element (or distribution) and of its covariance operator are equivalent.

We shall give some additional formulations for a covariance operator to be nondegenerate. Applying the Cauchy-Bunyakovksy inequality to the integral that defines R, we obtain the inequality

$$(Rf)(g) \leqslant \sqrt{(Rf)(f)}\,\sqrt{(Rg)(g)}\,,$$

which shows that $(Rf)(f) = 0$ if and only if $Rf = 0$. Therefore the fact that R is nondegenerate is equivalent to the fact that the mapping $R: X^* \to X^{**}$ is one to one.

Let $RX^* \subset X$ and let RX^* be everywhere dense in X (X is considered as a subspace of X^{**}). Then, the operator R is nondegenerate. In fact, if $Rf_0 = 0$, then $f_0(Rf) = f(Rf_0) = 0$ for all f and therefore $f_0 = 0$. Conversely, if RX^* is not everywhere dense in X, then according to the Hahn–Banach theorem, there exists an $f_0 \in X^*$, such that $f_0 \neq 0$ and $f_0(Rf) = 0$ for all f. But then $f(Rf_0) = 0$ for all f, and therefore $Rf_0 = 0$, which shows that R is degenerate.

Therefore if $RX^* \subset X$, then the fact that R is nondegenerate is equivalent to the following property: The closure in X^{**} of the set RX^* contains X (recall that the natural embedding of X into X^{**} is an isometry, and thus the norm in X is equal to the norm induced from X^{**}).

4.3.6

In the last two subsections we shall make two short additional remarks.

In some applications the assumption that the Banach space X is real is not satisfactory, creating a need to consider the complex case. Although all our results were obtained for real spaces, there is no difficulty in generalizing and applying them to the complex case. We shall only give the definition of the covariance operator in the complex case. Assume that the expectation is zero. Then the expression that defines the covariance operator of the random element x will be written in the form (for more details see reference 47)

$$(Rf)(g) = Eg(x(\omega))\overline{f(x(\omega))}, \qquad f, g \in X^*;$$

and it is obvious that instead of symmetry the operator R has the Hermitian property: $(Rf)(g) = \overline{(Rg)(f)}$ and instead of linearity it has the property of *conjugate linearity* (ther term *sesquilinearity* is also used): $R(\alpha f + \beta g) = \bar{\alpha} Rf + \bar{\beta} Rg$. The properties of nonnegativity and boundedness of the operator R remain unchanged. The first is obvious; the second can be established in the same manner as for the real case.

Among the other necessary changes in the complex case, it is worth noting that the lemma of Subsection 4.3.2 gives the factorization $R = A^*IA$ instead of $R = A^*A$, where I now denotes the involution operator in the complex Hilbert space $L_2(\Omega, \mathfrak{B}(\Omega), P)$, through which the factorization occurs.

4.3.7

With each weakly second-order random element x one can associate a particular bounded linear mapping of the space X^* into the Hilbert space $H = L_2(\Omega, \mathfrak{B}(\Omega), P)$ (of complex or real functions, according to whether X is complex or real), assigning to the element $f \in X^*$ the function $f(x(\cdot)) \in H$. This fact naturally brings to mind that in general the bounded linear mapping $X^* \to H$ can be considered as a generalization of the notion of a random element of the second order.

For random elements generalized in this way the characteristic functional, expectation, and covariance operator can be determined similarly in the usual case. This can easily be done by taking Af instead of $f(x(\omega))$, where $A : X^* \to H$ denotes the generalized random element. We shall not give explicit expressions. We remark that only under the assumption that the expectation is zero, the covariance operator R of the generalized random element A can easily be connected with the mapping given by the operator A. It is $R = A^*IA$ in the complex case, and in the real case it is $R = A^*A$.

One may also introduce the notion of a generalized Gaussian random element, calling it the mapping $A : X^* \to H$, which maps the space X^* into the subspace $H_1 \subset H$ consisting of Gaussian random variables. If the generalized Gaussian random element A is determined by an ordinary random element x; that is, if $Af = f(x(\cdot))$ for all $f \in X^*$, then it is obvious that x is a Gaussian random element. But then its covariance operator A^*A has to have some special properties. For example, if X is a real separable Hilbert space, then the operator A^*A should be an S-operator (this condition is also sufficient; one should take as x the Gaussian random element with zero mean and the covariance operatorA^*A). From this it follows, in particular, that the class of generalized random elements is in reality an extension of the class of all ordinary (weakly second order) random elements. For some problems in the theory of probability distributions on linear spaces such an extension of the notion of a random element is very important.

4.4 The Characterization of Classes of Covariance Operators

4.4.1

Again let X be a real Banach space and $\mathfrak{L}$ the smallest σ-algebra with respect to which all the continuous linear functionals are measurable. Recall that $\mathfrak{L}$ is generated by the Borel cylindrical sets (or even half-spaces); and when X is separable, then $\mathfrak{L}$ coincides with the σ-algebra $\mathfrak{B}(X)$ of the strong Borel sets. In the general case $\mathfrak{L} \subset \mathfrak{B}(X)$.

In this subsection it is more convenient to use the notion of distributions rather than random elements. By a distribution on X we mean, as before, the normed measure on the measurable space $(X, \mathfrak{L})$. The distribution μ is of second order in the weak sense if $E_\mu f^2(x) < +\infty$ for all $f \in X^*$.

Recall also that $\mathfrak{R} = \mathfrak{R}(X)$ denotes the family of all bounded linear symmetrical nonnegative mappings $X^* \to X^{**}$. We now introduce two new notions. Define $\mathfrak{R}_1 = \mathfrak{R}_1(X)$ to be the class of all covariance operators of distributions on X, which are weakly second order, and let $\mathfrak{R}_2 = \mathfrak{R}_2(X)$ be the class of covariance operators of all Gaussian distributions on X. It is obvious that $\mathfrak{R}_2 \subset \mathfrak{R}_1$. Furthermore, $\mathfrak{R}_1 \subset \mathfrak{R}$, as shown in Section 4.3. Therefore for each space X the following relations hold.

$$\mathfrak{R}_2(X) \subset \mathfrak{R}_1(X) \subset \mathfrak{R}(X).$$

If the space X is finite dimensional, then $\mathfrak{R}_2 = \mathfrak{R}$ and therefore all three classes coincide; and the conditions that determine $\mathfrak{R}$ characterize them completely. If X is infinite dimensional, then the equality $\mathfrak{R}_2 = \mathfrak{R}$ does not hold and the problem of characterization of the class $\mathfrak{R}_2$ becomes an interesting problem of describing all Gaussian distributions on X. It is also interesting to give the description of the class $\mathfrak{R}_1$ of all covariance operators and, in particular, to determine whether each operator $R \in \mathfrak{R}(X)$ is the covariance operator of some distribution on X; that is, whether the classes $\mathfrak{R}_1$ and $\mathfrak{R}$ coincide. These problems are the subject of the next subsections.

4.4.2

Let us start with the description of the class $\mathfrak{R}_1$. In the next three subsections we shall deal with nonreflexive spaces X and shall give different sufficient conditions for the images of a given element $f \in X^*$, or of the whole space X^* under the mapping given by the operator $R \in \mathfrak{R}_1(X)$, to belong to the space X.

To do this we shall again use Banach's theorem from Subsection 4.2.3, according to which the element Rf belongs to the space X if the expression $(Rf)(g)$ considered as a function of g is continuous in the X-topology of the space X^*. Because of the linearity of this functional, it is sufficient to check the continuity at zero. Let $\{g_n\}$ be an arbitrary sequence of bounded linear functionals in X, weakly convergent to zero; that is,

$$g_n(x) \to 0 \qquad \text{for all } x \in X. \tag{4.19}$$

According to the Cauchy–Bunyakovsky inequality, the following estimate holds.

$$[(Rf)(g_n)]^2 \leqslant (Rf)(f) \int_X g_n^2(x)\mu(dx), \tag{4.20}$$

where μ is the distribution with covariance operator R. We assume for brevity that the expectation of the distribution μ is zero.

Assume that μ is of second order in the strong sense ($E_\mu \|x\|^2 < +\infty$). Taking into account that the norms of the elements of the weakly convergent sequence of linear functionals are bounded, and using Lebesgue's well-known theorem about convergence of integrals, we obtain from (4.19) and (4.20) that $(Rf)(g_n) \to 0$ for each fixed element $f \in X^*$.

Thus we have proved the following theorem.

Theorem. *Let the space X be separable and let R be the covariance operator of the distribution μ. If μ is of second order in the strong sense, then $RX^* \subset X$.*

4.4.3

If the covariance operator R is given, then μ will denote, as before, one of the distributions having the covariance operator R (there exists, of course, an infinite set of distributions with the given covariance operator). Conversely, if the distribution μ is given, then R will denote its covariance operator (to a given distribution there corresponds, of course, only one covariance operator). It is obvious that by shifting a distribution by an arbitrary element $a \in X$ (i.e., by adding the constant a to the corresponding random element) the covariance operator remains unchanged. Therefore the expectation of the distribution may be chosen any way we wish. For convenience, from now on we shall assume that it is equal to zero.

Let f be an arbitrary given element in the space X^*. Repeating essentially the considerations of the proof of the theorem of Subsection 4.2.3 we can easily show, assuming that X is separable, that Rf belongs to the subspace $X \subset X^{**}$ if the following condition is satisfied: There exists a number $\epsilon > 0$, such that

$$\inf_{g \in X^*} \mu\{x : |f(x)g(x)| \geqslant \epsilon |E_\mu f(x)g(x)|\} > 0. \tag{4.21}$$

Now, we would like to give a condition of the same type (condition in the "weak form"), under which the image of the whole space X^* will be in X. Of course, one could simply require that condition (4.21) be satisfied for all $f \in X^*$. But by first applying the Cauchy–Bunyakovsky inequality, we obtain a more convenient condition.

Theorem. *Let the space X be separable. If for some $\epsilon > 0$*

$$\inf_{g \in X^*} \mu\{x : g^2(x) \geqslant \epsilon E_\mu g^2(x)\} > 0,$$

then $RX^* \subset X$.

PROOF. The proof of this theorem, similar to that of the theorem of Subsection 4.2.3, is based on the previously mentioned Banach theorem and on the following lemma, which gives a condition for interchanging the order of the limit operation and integration. □

Lemma. *Let $\{\xi_n\}$ be a sequence of nonnegative measurable functions defined on some measurable space $(\Omega, \mathfrak{B}(\Omega))$ with a finite measure P. Assume also that $\{\xi_n\}$ converges a.e. to zero. If for some $\epsilon > 0$*

$$\inf_n P\{\omega : \xi_n(\omega) \geqslant \epsilon E\xi_n(\omega)\} > 0,$$

then $E\xi_n \to 0$.

PROOF. The proof of this lemma follows from the proof of the theorem of Subsection 4.2.3. □

4.4.4

The probability that the square of a Gaussian random variable with zero mean is not less than its variance is a positive number that does not depend on the variance. Therefore for all Gaussian distributions on X the assumptions of the previous theorem are satisfied, and hence we have the following theorem.

Theorem. *Let the space X be separable. If R is the covariance operator of a Gaussian distribution, then $RX^* \subset X$.*

REMARK. The operators in the class $\mathfrak{R}(X)$ do not generally have this property. The simplest example is the operator $R \in \mathfrak{R}(c_0)$ whose matrix in the natural basis consists of ones only.

4.4.5

We shall now consider the following problem. Does the class $\mathfrak{R}(X)$ consist only of covariance operators, that is, do $\mathfrak{R}$ and $\mathfrak{R}_1$ coincide?

First, consider the case where X is an arbitrary separable Hilbert space H. Note, as an intermediate auxiliary result, that on this space there exists a distribution (and therefore also a random element) with the unit covariance operator. It can explicitly be shown that one of the distributions with the unit covariance operator will be obtained by taking masses $p_k/2$ at the points $\pm a_k e_k$, where e_k is a basis in H and

$$p_k > 0, \qquad \sum_{k=1}^{\infty} p_k = 1, \qquad a_k^2 p_k = 1, \quad (k = 1, 2, \ldots).$$

Let R be a given operator from the class $\mathfrak{R}(H)$, and let $e(\cdot)$ be one of the random elements with the unit covariance operator (e.g., the identity transformation in the probability space just constructed). One of the properties of the covariance operators formulated in Subsection 4.3.4 shows that the covariance operator of the distribution of the random element $x(\cdot) = R^{1/2}e(\cdot)$ is just R.

Therefore for each operator $R \in \mathfrak{R}(H)$ we can construct a distribution on H, which has the covariance operator R. Thus we have proved the following theorem.

Theorem. $\mathfrak{R}_1(H) = \mathfrak{R}(H)$.

4.4.6

One of the properties of covariance operators shows that if B is a bounded linear mapping $H \to X$, then the random element $x(\cdot) = Be(\cdot)$ with values in X, will have its covariance operator equal to $B^{**}B^*$. Therefore if the operator $R \in \mathfrak{R}(X)$ admits the factorization $B^{**}B^*$, then $R \in \mathfrak{R}_1(X)$ (one of the distributions with the covariance operator R will be the distribution of the random element Be). In what cases is the factorization of operators of the class $\mathfrak{R}(X)$ needed here possible? Lemma of Subsection 4.3.2 shows that each operator from this class can be factored as $R = A^*A$. Therefore the problem reduces to: In what cases is the operator $A : X^* \to H$ conjugate to some operator $B : H \to X$? In addition, we need the intermediate Hilbert space to be separable (the random element with the unit covariance operator in H has been constructed under the assumption of separability). It is obvious that if the space X is reflexive, then the operator $A : X^* \to H$ is conjugate to the operator $A^* : H \to X$. Furthermore, if X is separable, then H is also separable (see remark of Subsection 4.3.2). Therefore we obtain the following generalization of the previous theorem.

Theorem. *If the space X is separable and reflexive, then $\mathfrak{R}_1(X) = \mathfrak{R}(X)$.*

REMARK. Let $R \in \mathfrak{R}(X)$. If $RX^* \subset X$, then it is obvious that $A^*H \subset X$, where $A^*A = R$. Therefore in this case the operator A is conjugate to A^* (X is not assumed to be reflexive). From this we obtain, as in the case where X is separable, that if $RX^* \subset X$, $R \in \mathfrak{R}(X)$, then $R \in \mathfrak{R}_1(X)$.

4.4.7

In Subsection 4.1.6 we noted that in the case $X = Y^*$ the characteristic functional of a distribution on the space X can be defined in the wide sense (i.e., as a functional in $X^* = Y^{**}$), as well as in the narrow sense (as a functional in Y), and that these two approaches essentially coincide. Similarly, the covariance operator can be determined either in the wide sense, as we have done, or in the narrow sense, considering the corresponding quadratic functional on the subspace Y and not on the whole space Y^{**}. In this particular approach the covariance operator would be determined not as an operator $R : Y^{**} \to Y^{***}$ but as an operator $R_0 : Y \to Y^*$. Using the same general facts from functional analysis as in Subsection 4.1.6, we can easily show that these two approaches coincide if the condition $RY^{**} \subset Y^*$ is satisfied (i.e., in the sense of the natural embedding of Y^* into Y^{**}).

Recall that this condition for covariance operators of Gaussian distributions is always satisfied in separable Banach spaces $X = Y^*$ (Theorem of Subsection 4.4.4).

4.4.8

We now come to the characterization of Gaussian covariance operators. Theorem of Subsection 4.4.4 gives a necessary condition for membership in the class $\mathfrak{R}_2(X)$. The complete characterization of this class is known only for the spaces $X = \ell_p$, $1 \leqslant p < \infty$, and follows from the theorem of Subsection 2.2.6. It can be formulated as follows.

Theorem. *The operator $R \in \mathfrak{R}(\ell_p)$, $1 \leqslant p < \infty$, belongs to the class $\mathfrak{R}_2(\ell_p)$ if and only if the following condition is satisfied.*

$$\sum_{k=1}^{\infty} r_{kk}^{p/2} < +\infty, \tag{4.22}$$

where $\|r_{ij}\|$ is the matrix of the operator R in the natural basis ($r_{ij} = (Re_i, e_j)$, $e_n = e_{n,k}$, $e_{n,n} = 0$ for $k \neq n$ and $e_{n,n} = 1$).

For $p = 2$ condition (4.22) means that the trace of the matrix of the operator R is finite. It is well known that the trace of a matrix does not depend on the basis in ℓ_2 and, in particular, it is equal to the sum of the eigenvalues of the corresponding operator R. Thus for $p = 2$ condition (4.22) can be formulated explicitly in operator terms, without using the matrix representation. This invariant property (the sum of the series of all eigenvalues being finite) for the operators from the class $\mathfrak{R}(\ell_2)$ is well known in analysis. These operators are called *nuclear operators*.

Further, it is also obvious that the last theorem for $p = 2$ holds not only for the space ℓ_2 but also for an arbitrary real separable Hilbert space H and gives the following complete characterization of the class $\mathfrak{R}_2(H)$: the operator $R \in \mathfrak{R}(H)$ belongs to the class $\mathfrak{R}_2(H)$ if and only if it is a nuclear operator.

4.4.9

The notion of nuclearity exists not only for the operators on Hilbert spaces but also for the operators acting between Banach spaces. Naturally, the following problem arises: Is nuclearity of an operator the additional property that allows us to select the class $\mathfrak{R}_2(X)$ from the whole class $\mathfrak{R}(X)$, not only for Hilbert spaces but also for the general case of Banach spaces X? The answer is negative. One can see this by examining the situation for the spaces ℓ_p, $p \neq 2$, using the theorem of Subsection 4.4.8.

However, we shall show that this selection property is connected with the notion of nuclearity in another sense, which is quite natural when applied to mappings of spaces into their conjugates. We first note that the mapping $R \in \mathfrak{R}(H)$ is called a *Hilbert–Schmidt operator* if the sum of the squares of all the eigenvalues is convergent. Recall that the product of two arbitrary

Hilbert–Schmidt operators is a nuclear operator; and, conversely, each nuclear operator is a product of two Hilbert–Schmidt operators. Note that the definition and basic properties of Hilbert–Schmidt operators and nuclear operators can be formulated for a broader class of operators than the nonnegative ones. The general case can easily be reduced to the case of operators from the class $\Re(H)$, because each bounded linear operator in H can be represented as a product of a unitary and a nonnegative operator (for a more detailed description of these notions and the proofs for the Hilbert space case, see Gelfand and Vilenkin [54], pp. 32–52).

Now, we shall give two definitions of nuclearity for operators of the class $\Re(X)$—of Hilbert–Schmidt operators $H \to X$ (and $X \to H$). We say that the operator $A : H \to X$ is a *Hilbert–Schmidt operator in the first sense* if the factorization $A = BT$ takes place, where T is a Hilbert–Schmidt mapping $H \to H$ and B is a bounded linear mapping $H \to X$. Similarly, we define the Hilbert–Schmidt mapping $X \to H$. Further, we say that $A : H \to X$ is a *Hilbert–Schmidt operator in the second sense* if CA is a Hilbert–Schmidt operator for all bounded linear mappings $C : X \to H$. Similarly, we define the Hilbert–Schmidt mapping $X \to H$.

The operator $R \in \Re(X)$ will be called a *nuclear operator in the first* (or correspondingly in the *second*) *sense*, if in its factorization $R = A^*A$ (see lemma of Subsection 4.3.2) the operator $A : X^* \to H$ is a Hilbert–Schmidt operator in the first (or correspondingly in the second) sense (then $A^* : H \to X^{**}$ will also be a Hilbert–Schmidt operator in the same sense).

Theorem. *Let the space X be separable and reflexive and let $R \in \Re(X)$. Nuclearity of R in the first sense is sufficient and in the second sense is necessary for $R \in \Re_2(X)$.*

PROOF. Assume that R is a nuclear operator in the first sense. Then, R can be represented in the form

$$R = A^*A = B^*TTB, \tag{4.23}$$

where B is a bounded linear mapping $X^* \to H$ and T is a Hilbert–Schmidt mapping $H \to H$. The operator $T^2 : H \to H$ is a nuclear operator because it is the product of two Hilbert–Schmidt operators. Furthermore, the intermediate Hilbert space H is separable because of the restrictions on the space X. Therefore, according to the theorem of Subsection 4.4.8 (with $p = 2$), $T^2 \in \Re_2(H)$; and hence there exists a Gaussian random element $h(\omega)$, $\omega \in \Omega$, in H, with the covariance operator T^2. Since X is reflexive, the operator B^* in the factorization (4.23) maps H into X. It is easy to show that the random element $B^*h(\cdot)$ with values in X will be Gaussian and have the covariance operator R. This proves that $R \in \Re_2(X)$. Conversely, let $R \in \Re_2(X)$ and let $x(\omega), \omega \in \Omega$, be a Gaussian random element with the covariance operator R. Let an arbitrary bounded operator $C : H \to X^*$ be given. The random element $C^*x(\cdot)$ with values in H will be Gaussian with the covariance operator C^*RC. Therefore this operator should be a nuclear operator in the ordinary sense (as an operator $H \to H$), and the factorization $R = A^*A$ shows that $(AC)^*AC$ is a nuclear operator for each $C: H \to X$. From this it follows that all $AC: H \to H$ are Hilbert–Schmidt operators, which means that $R = A^*A$ is a nuclear operator in the second sense.

Thus we have proved the theorem. Note that X has been assumed reflexive only in order to simplify the proof. □

The proof of this theorem shows, in particular, that if R is a nuclear operator in the first sense, then it is nuclear in the second sense, too. (This fact can also be shown explicitly without any restriction on X.) The converse assertion is true at least in the case when the space X is linearly homeomorphic to some Hilbert space. The proof follows easily from the definitions.

4.4.10

Finally, we shall briefly consider a topological way of describing the class $\mathfrak{R}_2(X)$ and note the connection with the problem of describing all Gaussian distributions in X by means of characteristic functionals.

An elementary consideration shows that the characteristic functional of a Gaussian distribution μ on an arbitrary space X has the form (we assume for simplification that the expectation is zero)

$$\chi(f;\, \mu) = \exp\{-\tfrac{1}{2}(Rf)(f)\}, \qquad f \in X^*, \tag{4.24}$$

where R is the covariance operator of the distribution μ. It is obvious that $R \in \mathfrak{R}_2(X)$ (from the definition of this class). Conversely, if $R \in \mathfrak{R}_2(X)$, then (4.24) is a characteristic functional (of the Gaussian distribution with expectation zero and covariance operator R).

Therefore (4.24) is the Fourier transform of some (Gaussian) distribution on X if and only if $R \in \mathfrak{R}_2(X)$.

The condition for the functional (4.24) to be a characteristic functional (i.e., for R to belong to the class $\mathfrak{R}_2(X)$) can be given in topological terms as the continuity of $\chi(f;\, \mu)$, that is, as the continuity of $(Rf)(f)$ in some topology $\mathfrak{T}$ which we shall now define.

The functional $p(f) = [(Rf)(f)]^{1/2}$ is for each $R \in \mathfrak{R}(X)$ a prenorm (seminorm) in X^*, that is, $p(f) \geqslant 0$, $p(\alpha f) = |\alpha| p(f)$, and $p(f+g) \leqslant p(f) + p(g)$ (the first two properties are obvious, the third follows from the inequality $|(Rf)(g)| \leqslant p(f)p(g)$, which can be established similarly to the Cauchy–Bunyakovsky inequality for the inner product).

Let $\mathfrak{P}$ be a system of prenorms that satisfy the following two conditions:

a. If $p_1, p_2, \ldots, p_n$ belong to $\mathfrak{P}$ and $\epsilon_1, \epsilon_2, \ldots, \epsilon_n$ are positive numbers, then $\epsilon_1 p_1 + \epsilon_2 p_2 \cdots + \epsilon_n p_n \in \mathfrak{P}$,
b. For an arbitrary nonzero element $f_0 \in X^*$ there exists a prenorm $p_0 \in \mathfrak{P}$ such that $p_0(f_0) > 0$.

It is easy to check explicitly that each such system of prenorms generates in X^* a locally convex Hausdorff topology consistent with the linear structure, provided that all sets of the form $\{f: p(f) < 1\}, p \in \mathfrak{P}$ are taken as the basis of neighborhoods of zero. Particularly, the system of prenorms $[(Rf)(f)]^{1/2}$ created by the whole family of operators from the class $\mathfrak{R}_2 \subset \mathfrak{R}$ obviously satisfies these conditions and generates a topology that will be the $\mathfrak{T}$-topology. Indeed, the necessity of the continuity of $(Rf)(f)$ for R to belong to the class $\mathfrak{R}_2$ is obvious. We shall show sufficiency. Let $R \subset \mathfrak{R}$ and let the functional $(Rf)(f)$ be continuous in the $\mathfrak{T}$-topology. Then, for an arbitrary $\epsilon > 0$, particularly for $\epsilon = 1$, there exists an operator $\tilde{R} \in \mathfrak{R}_2$ such that the condition $(\tilde{R}f) < 1$ implies $(Rf)(f) < 1$. But because the functionals are quadratic this is equivalent to the condition

$$(Rf)(f) \leqslant (\tilde{R}f)(f) \qquad \text{for all} \quad f \in X^*. \tag{4.25}$$

Thus we have only to show that the class $\mathfrak{R}_2(X)$ includes, together with each operator $\tilde{R}$, all the operators from $\mathfrak{R}(X)$ that are comparable with $\tilde{R}$ and not greater than $\tilde{R}$ (in the sense of relation (4.25)). This fact follows easily from the following result, which is due to A. Badrikian, A. Tortrat [30]: If χ is a characteristic functional, χ' is a nonnegative definite functional, $\chi'(0) = 1$, and $|1 - \chi'| \leqslant |1 - \chi|$ everywhere, then χ is also a characteristic functional. This result is proved for Borel distributions. Therefore for the cases where $\mathfrak{L} = \mathfrak{B}(X)$ (particularly for X separable), the theorem on the topological nature of the conditions determining the class $\mathfrak{R}_2(X)$ is proved.

This theorem can be interpreted as a theorem on the topological nature of the conditions for extending a Gaussian weak distribution to a measure. The topological nature of these conditions is noted in the paper by V. N. Sudakov and A. M. Vershik [50].

References

1. Kolmogorov, A. 1935. La transformation da Laplace dans les espaces linéaires. *C. R. Acad. Sci. Paris* 200: 1717–1718.
2. Fréchet, M. 1951. Generalisation de la loi de probabilité de Laplace. *Ann. Inst. H. Poincaré* 12: 215–310.
3. Mourier, E. 1953. Éléments aléatoires dans un espace de Banach. *Ann. Inst. H. Poincaré* 13: 161–244.
4. Prohorov, Ju. V. 1956. Convergence of stochastic processes and limit theorems in the theory of probability. *Theory of Probability and Its Applications* 1: 157–214.
5. Sazonov, V. V. 1958. A remark on characteristic functionals. *Theory of Probability and Its Applications* 3: 188–192.
6. Minlos, P. A. 1959. Generalized random processes and their extension to a measure (Russian). *Trudy Moskovskogo Matematitsheskogo Obshtshestva* 8: 497–518.
7. Kolmogorov, A. N. 1959. A note on the papers by P. A. Minlos and V. V. Sazonov. *Theory of Probability and Its Applications* 4: 221–223.
8. Vakhania, N. N. 1965. On characteristic functionals of random sequences (Russian). *Trudy VC AN GSSR* 5: 5–32.
9. Vakhania, N. N. 1965. Sur les répartitions de probabilités dans les espaces de suites numériques. *C. R. Acad. Sci. Paris* 260: 1560–1562.
10. Prohorov, Yu. V. 1961. The method of characteristic functionals. *Proc. 4th Berkeley Symposium on Math. Stat. and Prob.* Berkeley and Los Angeles: University of California Press. pp. 403–419.
11. Dunford, N. and Schwartz, J. 1966. *Linear Operators, Part* 1. New York: Interscience Publishers.
12. Bourbaki, N. 1955. *Espaces Vectoriels Topologiques*. Paris: Hermann.
13. Loéve, M. 1963. *Probability Theory*, 3rd ed. New York: Van Nostrand.
14. Halmos, P. R. 1950. *Measure Theory*. New York: Van Nostrand.

15. Schwartz, L. 1968. Démonstration de deux lemmes sur les probabiltés cylindriques. *C. R. Acad. Sci. Paris* 266: 50–52.

16. Sazonov, V. V. 1962. On characteristic functionals (Russian). *Trudy VI-go Vsesoyuznogo Soveshtshania po Teorii Verojatnosti i Matehatitsheskoj Statistike*. 1960, *Vilnius*. Gosudarstvenoje Izadelstvo Polititsheskoj i Nautshnoj Literatury Lit. SSR, 1962, pp. 449–454.

17. Prohorov, Yu. V. and Sazonov, V. V. 1961. Some results associated with Bochner's theorem. *Theory of Probability and Applications* 6: 82–87.

18. Schwartz, L. 1967. Extension du théoréme de Sazonov-Minlos á des cas non hilbertiens. *C. R. Acad. Sci. Paris* 265: 832–834.

19. Schwartz, L. 1968. Réciproque du théoréme de Sazonov-Minlos dans des cas non hilbertiens, *C. R. Acad. Sci. Paris* 266: 7–9.

20. Kwapien, S. 1968. Complement au théoréme de Sazonov-Minlos, *C. R. Acad. Sci. Paris* 267: 698–700.

21. Fortet, R. et Mourier, E. 1955. Les fonctions aléatoires comme éléments aleatoires dans les espaces de Banach. *Studia Math* 15: 62–79.

22. Varadarajan, V. S. 1961. Convergence of stochastic processes. *Bull. Amer. Math Soc.* 67: 276–280.

23. Vakhania, N.N. 1964. On normal distributions in l_p spaces. *Theory of Probability and Its Applications* 9: 665.

24. Vakhania, N. 1965. Sur une propriété des répartitions normales de probabilites dans les espaces l_p ($1 \leqslant p < \infty$) et H. *C. R. Acad. Sci. Paris* 260: 1334–1336.

25. Vakhania, N. N. 1966. On non-degenerate probability distributions in $l_p(1 \leqslant p < \infty)$ spaces. *Theory of Probability and Its Applications* 11: 463–467.

26. Vakhania, N. N. 1967. On some probabilistic problems for a one-dimensional heat equation. *Theory of Probability and Its Applications* 12: 666–667.

27. Fortet, R. 1954. Normalverteilte Zufallselemente in Banachschen Räumen, *Bericht über die Tagung Wahrscheinlichkeitsrechnung und Math. Stat. in Berlin*, Berlin. pp. 29–35.

28. Shilov, G. E. and Fan Dick Tin, 1967. *Integral, Measure and Derivatives on Linear Spaces* (Russian). *Nauka*, Moscow.

29. Vershik, A. M. 1964. The general theory of Gaussian measures in linear spaces (Russian). *Uspehy Matematitsheskih Nauk* 19: 210–212.

30. Tortrat, A. 1965. Lois indéfiniment divisibles ($\mu \in I$) dans un group topologique abélien métrisable X. Cas des espaces vectoriels. *C. R. Acad. Sci. Paris* 261: 4973–4975.

31. Whittaker, E. T. and Watson, J. N. 1958. *A Course in Modern Analysis*, 4th ed. Cambridge: Cambridge University Press.

32. Beilinson, A. A. 1964. On solution of probabilistic unbalanced problem for distributed systems. *Theory of Probability and Its Applications* 9: 469–472.

33. Feller, W. 1960. *Introduction to Probability Theory and Its Applications*, *Vol.* 1, 2nd ed. New York: Wiley.

34. Sazonov, V. V. 1968. On the multi-dimensional central limit theorem. *Sankhya* 30: 181–204.

35. Kandelaki, N. P. 1965. On some limit theorems in Hilbert space (Russian). *Trudy VC AN GSSR* 5: 46–55.

36. Sazonov, V. V. 1968. On ω^2 criterion. *Sankhya* 30: 205–210.

37. Daletskiĭ, Yu. L. 1967. Infinite-dimensional elliptic operators and related parabolic equations (Russian). *Uspekhi Matematitsheskih Nauk* 22: 4–54.

38. Vakhania, N. N. and Kandelaki, N. P. 1968. On the estimation of the convergence rate in the multi-dimensional central limit theorem (Russian). *Soobshtshenia AN GSSR* 50: 273–276.

39. Vakhania, N. N. and Kandelaki, N. P. 1969. On the estimation of the convergence rate in the central limit theorem in Hilbert space (Russian). *Trudy VC AN GSSR* 9: 150–160.

40. Vakhania, N. N. and Kandelaki, N. P. 1968. On the distribution of the inner product of Gaussian random vectors (Russian). *Soobshtshenia AN GSSR* 50: 535–540.

41. Vakhania, N. N. and Kandelaki, N. P. 1967. The stochastic integral for operator valued functions. *Theory of Probability and Its Applications* 12: 525–528.

42. Prohorov, Yu. V. and Rozanov, Yu. A. 1969. *Probability Theory* (translated from the Russian). New York: Springer-Verlag.

43. Ibraginov, I. A. and Linnik, Yu. V. 1971. *Independent and Stationary Sequences of Random Variables* (translated from the Russian). Gronigen: Wolters-Noordhoff.

44. Ibraginov, I. A. 1963. On estimation of the spectral function of the stationary Gaussian process. *Theory of Probability and Its Applications* 8: 366–401.

45. Gihman, I. I. and Skorohod, A. V. 1969. *Introduction to the Theory of Random Processes*. (translated from the Russian). Philadelphia: Saunders.

46. Vakhania, N.N. 1968. On an existence condition of the Pettis integral. *Studia Math* 29: 243–248.

47. Vakhania, N. N. 1968. The covariance operator of a probability distribution on a Banach space (Russian). *Soobshtshenia AN GSSR* 51: 35–40.

48. Vakhania, N. N. 1969. On the covariance of random elements in linear spaces (Russian). *Soobshtshenia AN GSSR* 53: 17–20.

49. Le Cam, L. 1957. Convergence in distribution of stochastic processes. *Univ. Calif. Publs. Statist.* 2: 207–236.

50. Sudakov, V. N. and Vershik, A. M. 1962. Topological problems of the measure theory in linear spaces (Russian). *Uspehy Matematitsheskih Nauk* 17: 217–219.

51. Grenander, U. 1963. *Probabilities on Algebraic Structures*. New York: Wiley.

52. Hille, E. and Phillips, R. 1957. *Functional Analysis and Semigroups*. American Mathematical Society. Providence, R.I.

53. Banach, S. 1932. *Theorie des opérations linéaries*. Warsaw: Funduszu Kultury Narodowej.

54. Gelfand, I. M. and Vilenkin, N. Ya. 1964. *Generalized Functions, Volume* 4, *Applications of Harmonic Analysis* (translated from the Russian). New York: Academic Press.

Author Index

Subject Index